Experiments and Research on Fracture Behaviors of Dam Concrete

Zhifang Zhao, Hougui Zhou, Zhigang Zhao

中国建筑工业出版社
China Architecture & Building Press

图书在版编目（CIP）数据

大坝混凝土断裂性能试验与研究=Experiments and Research on Fracture Behaviors of Dam Concrete：英文/赵志方，周厚贵，赵志刚著. —北京：中国建筑工业出版社，2017.12
ISBN 978-7-112-21554-6

Ⅰ.①大… Ⅱ.①赵…②周…③赵… Ⅲ.①混凝土坝—大坝—断裂性能—试验研究—英文 Ⅳ.①TV642

中国版本图书馆CIP数据核字（2017）第289365号

All rights reserved. No part of this publication may be reproduced or transmitted in any form or by any means, electronic or mechanical, including photocopying, recording or any information storage or retrieval system, without prior permission in writing form the publishers.
@ 2017 China Architecture & Building Press, No.9, Sanlihe Rd., 100037, Beijing, China

Executive Editor：Duan Ning, Wang Liyao

Experiments and Research on Fracture Behaviors of Dam Concrete
Zhifang Zhao, Hougui Zhou, Zhigang Zhao

*

中国建筑工业出版社出版、发行（北京海淀三里河路9号）
各地新华书店、建筑书店经销
北京科地亚盟排版公司制版
北京圣夫亚美印刷有限公司印刷

*

开本：787×1092毫米 1/16 印张：13¼ 字数：402千字
2017年12月第一版 2017年12月第一次印刷
定价：**79.00元**
ISBN 978-7-112-21554-6
（31187）

版权所有 翻印必究
如有印装质量问题，可寄本社退换
（邮政编码100037）

序 I

  大坝是国家的重要基础设施之一。在中国乃至世界的大坝建设中，混凝土坝是使用最多、最广的一种坝型，其主要优点是承载能力强、使用寿命长、断面相对较小、可经受水流漫顶等。但在混凝土坝建设中，早期处于引领地位的欧美等一些国家曾发生过坝体混凝土开裂漏水、溃坝等重大事故，造成工程功亏一篑和巨大生命财产损失。这使得大坝混凝土开裂成为导致其耐久性和抗震性能大幅降低、使用寿命急剧缩短、严重影响其正常运行和发挥工程效益的世界级难题。因此，施工期混凝土的开裂控制和运行期的裂缝控制是大坝混凝土工程的重大关键技术问题之一。

  2002年3月，正值长江三峡水利枢纽二期大坝工程施工进入到高峰期，泄洪坝段混凝土局部出现了一些裂缝，为探讨长江三峡泄洪坝混凝土施工期裂缝成因和预防及处理对策，由清华大学、中国葛洲坝集团公司和烟台大学三方合作启动了"大体积混凝土裂缝仿真断裂分析研究"课题。该课题紧密结合当时正在施工的主体大坝的原材料及浇筑工艺，开展了大量的室内、现场试验和模拟计算研究，并以此为基础，持续开展了长达15年之久的国内外协同攻关和相关研究工作。

  研究工作在清华大学水电工程系和中国葛洲坝集团公司试验中心两地同步展开，并进行了广泛的考察调研、查新，较大规模地制备了全级配混凝土试件，系统实施了三峡大坝泄洪坝段混凝土断裂力学性能试验。试验用混凝土试件在三峡施工现场仿真制作，共浇筑试件377件，混凝土方量达23.1m³。试验工作在清华大学水电工程系高坝大型结构国家专业实验室与宜昌葛洲坝集团公司试验中心分别进行，采用了先进的测试技术，自主研制了关键试验装置，在大坝混凝土断裂性能方面取得重要技术创新研究成果。

  在三峡大坝混凝土断裂性能试验数据成果的基础上，还与美国西北大学先进水泥基材料研究中心（ACBM）主任、美国工程院院士、中国工程院外籍院士Professor Surendra. P. Shah，德国莱比锡应用技术大学教授、混凝土结构和材料专家Professor Volker Slowik等，对三峡大坝混凝土断裂性能的计算机模拟开展了国际合作和系统研究。其研究成果揭示了三峡大坝混凝土断裂机理，并得出了系统的断裂参数，使得该研究成果具有更深的意义和更广泛的应用价值。

  作为时任中国长江三峡工程开发总公司的总工程师，我一直关注该项课题的研究进展，先后考察了宜昌葛洲坝试验中心的试件场区、试验装备和试验情况，听取了研究团队的研究工作方案介绍，共同探讨了研究和试验过程中可能出现的问题和解决措施，提出了全面做好此项试验研究的相关意见和建议。我与研究团队的主要负责人也是此书的主要作者赵志方博士和周厚贵总工程师是在三峡工程的建设之中相识的，他们的刻苦学习与勇于攻关的精神和勤奋努力与扎实工作的作风给我留下了深刻的印象。

  本书正是对上述课题试验和研究成果的全面系统的总结，其主要创新点是：提出了大坝混凝土软化曲线简便适用的确定方法，并研发了大坝混凝土软化曲线的专用计算软件；利用裂缝扩展路径上应力集中和应力释放的现象，采用新颖的电测技术，观测断裂全过程中裂缝起裂、稳定扩展和失稳破坏的规律，提出了起裂点和裂缝扩展长度的测试方法；通过大坝混凝土断裂机理的揭示，利用软化曲线从机理上解释断裂能的尺寸效应；发现了大坝混凝土双K断裂参数无尺寸依赖性，从而完善了大坝混凝土的断裂准则；优化了大坝混凝土配合比，形成了一整套有效控制大坝裂缝的施工技术和工艺。

相信此书的出版能为广大的坝工建设者提供一个大坝混凝土防裂、控裂乃至建造无裂缝大坝的工程理论和工程实践的典型案例，并对今后大坝工程建设技术进步产生较大的促进作用。

中国工程院院士

2017年2月17日

Preface I

Dams are vital infrastructures in a country. As for dam construction in China and throughout the world, concrete is the most widely used material in their construction. The main advantages of concrete dam are its strong carrying capacity, long life, small cross-section, and good resistance to water overtopping, etc. However, dam concrete cracking, leakage, failure and other major accidents which occurred during their construction in Europe and the United States were commonplace in early development. This often resulted in the complete ruin of these dams and the loss of peoples lives and property. Thus, dam concrete cracking is a worldwide problem resulting in a significant reduction in durability, seismic performance and service life and has a serious impact on their normal operation and benefits. Consequently, concrete cracking control during the construction period and crack control during the operation period is one of the key technical problems in concrete dam engineering.

In March 2002, when the second phase of dam construction of the Yangtze River Three Gorges Project entered construction peak, some cracks were found in the concrete of the flood discharge dam. To explore the cause of this cracking, crack prevention and treatment countermeasures during the construction period of the Three Gorges flood discharge dam, Tsinghua University, China Gezhouba (Group) Corporation (CGGC) and Yantai University, all located in China, jointly launched the project Fracture Simulation Analysis of Mass Concrete. Using the raw materials and pouring process of the main dam under construction, many laboratory tests, field tests and simulation studies were carried out. Based on this, domestic and international collaborative research and related research work has been pursued for 15 years to try to solve the key technical problems.

The research work was carried out synchronously at the Department of Hydraulic Engineering of Tsinghua University and Test Center of CGGC. The research group made investigation and research, novelty search, prepared the fully graded concrete specimens on a large scale, systematically conducted fracture experiments of Three Gorges Dam concrete. Concrete specimens were prepared at the construction site of the Three Gorges Dam. A total of 377 specimens, with concrete volumes of up to 23.1 cubic meters were produced. The tests were performed in the National Professional Laboratory of High-dam Large-scale Structure of Tsinghua University and Test Center of CGGC, respectively. Advanced testing technology was employed, key test equipment was developed independently and important innovation achievements were made about fracture behaviors of dam concrete.

On the basis of fracture experiments data of the Three Gorges Dam concrete, they also worked with Professor S. P. Shah and Professor Volker Slowik, etc. Professor Shah, the director of the Advanced Cement-based Materials Research Center at Northwestern University, Evanston, Illinois in the United States, is a member of National Academy of Engineering of the United States and a foreign member of the Chinese Academy of Engineering. Professor Volker Slowik is an expert of concrete structures and materials worked at Hochschule für Technik, Wirtschaft und Kultur in Leipzig, Germany. They helped to carry out further international cooperation and systematic research on the computer simulation of the concrete fracture behavior of the Three Gorges Dam concrete. The research results revealed fracture mechanisms of the Three Gorges Dam concrete and obtained systematic fracture parameters, which gave the research results more meaningful and wider application values.

As the chief engineer of the China Three Gorges Corporation, I have been paying close attention to the research progress of the project. I have observed and studied the test specimen area, test equipment and test situation of Test Center of CGGC in Yichang, listened to presentations by the research group, discussed the problems and solutions that might occur in the research and testing process with them, and put forward relevant opinions and suggestions to comprehensively carry out this experimental study. I met the directors of the research team, also the main authors of this book, Dr. Zhifang Zhao and chief engineer Hougui Zhou, during the construction of the Three Gorges Dam project. I was very impressed by their hard work, study of key technical problems, diligence and professionalism.

This book is a comprehensive and systematic summary of the above-mentioned research achievements. The main innovations which resulted are as follows: The simple and applicable method for determining the softening curve of dam concrete was proposed and the corresponding specialized software was developed. Based on the phenomenon of stress concentration and stress release on the crack propagation path, the rule of crack initiation, stable propagation and unstable failure was observed by employing the innovative electrical measuring technology. The methods for testing crack initiation and crack propagation length were also proposed. The size effect of fracture energy was explained by revealing the dam concrete fracture mechanism from the viewpoint of the softening curve. It was discovered that the double-K fracture parameters of dam concrete were size independent, thereby improving the fracture criterion of dam concrete. The dam concrete mix proportions were optimized and formed a set of construction technology to effectively control dam cracks.

I believe that the publication of this book will provide most of the dam builders with a typical case of engineering theory and engineering practice for dam concrete crack mitigation and the control and even construction of non-crack dams. It will definitely play an important role in promoting the progress of dam construction technology in the future.

Chaoran Zhang
Member of the Chinese Academy of Engineering

February 17, 2017

Preface II

Cracking of concrete dams is one of the common problems during the construction period and operation period of the hydropower station. Cracks will greatly reduce a dam's durability and seismic performance. Severe cracking can shorten its service life. In this book, the process of cracking in dams is explained and explored by employing theoretical fracture mechanics. With the aid of fracture mechanics, the author has provided a tool to predict and control cracking of dams and other massive concrete structures.

Based on the actual needs of dam construction in China, in March 2002 a project was initiated to explore the causes, prevention and control of cracks of the flood discharge dam of the Three Gorges Project at Yangtze River. It was set up during the dam construction period, under the guidance of chief engineer Mr. Hougui Zhou and Professor Qingbin Li. The co-author Dr. and Professor Zhifang Zhao presided over the implementation of testing and research task Fracture Simulation Analysis of Mass Concrete. This project was jointly carried out by Tsinghua University, China Gezhouba (Group) Corporation and Yantai University. It was comprised of conducting large-scale fracture tests of dam concrete. Testing of large concrete specimens for evaluating fracture parameters was never done before and posed many technical challenges.

Three types of fracture test specimens, namely, the uniaxial tension test specimen, three-point bending notched beam, and wedge-splitting test specimens were employed. Specimens made with normal size aggregates as used in building and road construction, as well as specimens with large aggregates used in dam construction were tested. To compare these two sets of tests, specimens were made with wet-screening concrete. The maximum specimen size of wedge-splitting test specimens were: 1.2m long, 1.2m high and 0.25m thick. The fracture parameters of various types of dam concrete were measured and analyzed systematically. It was found that the double-K fracture parameters of dam concrete proposed by Professor Xu and studied by Professor Zhao can be used as a valid fracture parameters. The tested double-K fracture parameters of dam concrete were used to accurately simulate concrete cracks in the Three Gorges Dam. This study provided a method for the safety assessment and crack control of the Three Gorges Dam. In addition, the study helped optimization of concrete mix proportions.

I met Dr. Zhao at the 11th International Fracture Conference held in Turin, Italy, in March 2005. Her presentation: The *Experiments for Determining the Double-K Fracture Parameters of Concrete of the Three Gorges Dam* attracted my attention and received the praise of the conference Chair Professor A. Carpinteri. She introduced her comprehensive and systematic fracture experimental study of the Three Gorges Dam concrete. Since fracture mechanics of concrete was one of the important research directions of NSF's Advanced Cement-Based Materials Research Center (ACBM Center), I invited her to come to the United States in 2007 to pursue research under my supervision. During her visit, we jointly completed the project "Computer Simulation of Fracture Behaviors of Dam Concrete." Utilizing the comprehensive and systematic fracture tests of dam concrete of the Three Gorges Dam, we conducted the inverse analysis and finite element simulation of the test data. We proposed a simple and suitable method for determining the softening curve of dam concrete. The softening curve can explain the measured size effect of fracture energy. Dr. Zhao has used modern digital technology to explore this world-class engineering problem. Research results were published in international journals and have been widely cited by scholars.

During her stay at the ACBM Center at Northwestern University in America, Dr. Zhao carried out fruitful research in concrete fracture mechanics combining theory, test techniques, and computer simulations of dam concrete fracture. As an outstanding and promising scholar, in 2013 I invited Dr. Zhao to come to ACBM center to participate in a Sino-US cooperation project on "High Volume Mineral Admixture Mass Concrete." I recommended that Dr. Zhao visit University of Illinois at Urbana-Champaign to cooperate with Professor David A. Lange, who has carried out temperature-stress tests of concrete in the early stage. I also suggested that she visit Iowa State University to cooperate with Professor Kejin Wang, who is an expert on fly ash mass concrete. Dr. Zhao is keen on continuing her study of crack prevention and crack control of dams at her own Tsinghua University and Zhejiang University of Technology as well with active international collaboration.

This book, EXPERIMENTS AND RESEARCH ON FRACTURE BEHAVIORS OF DAM CONCRETE, co-authored by eminent scholars: Dr. Zhao and Zhou with its systematic study of the fracture behaviors of dam concrete will effectively guide dam safety assessment, crack control, and dam safety. It will provide a useful reference for college teachers and students as well as the engineering and technical personnel who are engaged in scientific research, design, construction, and management of dam concrete.

Surendra P. Shah
Walter P. Murphy Professor of Civil Engineering (emeritus),
Northwestern University, Evanston, IL USA
Member of National Academy of Engineering, USA
Foreign Member of Chinese Academy of Engineering,
Foreign Member of Indian National Academy of Engineering

Summary

Fracture of dam concrete is one of the main hidden dangers affecting and endangering dam safety during their construction and operation. To master and reveal the fracture characteristics and mechanism of mass concrete, while seeking to prevent and control cracks in dam, it is necessary to carry out concrete fracture experiments of large size and fully graded aggregate concrete specimens, together with corresponding tests on the physical and mechanical properties of concrete based on large-scale project. In 2002 the second phase of the Three Gorges project was at its peak, which resulted in a rare engineering opportunity for this important research. Thus Tsinghua University, China Gezhouba (Group) Corporation (CGGC) and Yantai University entered into a cooperative research agreement for "Fracture Simulation Analysis of Mass Concrete", they further agreed to study the key technical problems involved in the cracking and fracture of mass concrete structures to provide the basis for engineering construction and operations.

After the research program was begun, it took two and a half years to complete the preliminary experiments and research necessary to obtain the initial achievements. The experiments were simultaneously conducted at the National Professional Laboratory of High-dam Large-scale Structure of Tsinghua University and Test Center of China Gezhouba (Group) Corporation. At a later stage, research work also was conducted in the United States and Germany, with collaboration of academicians, professors or leading experts from Northwestern University, Evanston, IL, University of Illinois, Urbana-Champaign, IL, and Iowa State University, Ames, IA in America and the Hochschule für Technik, Wirtschaft und Kultur Leipzig in Germany, resulting in the research achievements set forth in this book.

This research initially determined the double-K fracture parameters of dam concrete of the Three Gorges Project. Its related achievements won the Hubei Science and Technology Progress Award in 2005. In the same year, the *Norm for Fracture Test of Hydraulic Concrete* (DL/T 5332-2005) was jointly edited by CGGC based on these fracture experimental test data becoming the basis for support of the national standard in China.

The research achievements have been well-received by Professor Alberto Carpinteri, Professor of science in construction at the Polytechnic University of Turin who was director of the Istituto Nazionale di Ricerca Metrologica (INRiM) in Turin, Chair of the 11th International Fracture Conference held in Turin of Italy in 2005, as well as have been fully affirmed by distinguished Professor Surendra. P. Shah, the Walter P. Murphy Emeritus Professor of Civil and Environmental Engineering at Northwestern University. He invited a research group member to cooperate with Northwesterns' Advanced Cement-based Materials (ACBM) Center to further study the computer simulation of the Three Gorges Dam concrete fracture behavior. These achievements have been published in well-known international journals, such as *Cement and Concrete Research*.

This book reveals that by employing uniaxial test, three-point bending notched beam test and wedge-splitting test, the fracture parameters and fracture mechanism of dam concrete and wet-screening concrete were investigated on which the effects of mix proportions, specimen sizes and coarse aggregate sizes are considered. Accordingly, the fracture behaviors of the Three Gorges Dam concrete were achieved, as well as helpful suggestions and solutions to mitigate and prevent cracking of dam concrete in the future. Its sixteen chapters include various topics on dam concrete: state-of-the-art review of concrete softening curve; determination of softening curve of dam concrete by inverse analysis based on cracking strength criterion, and comparison between this method with the Direct Tension (DT) method; research on fracture behaviors of dam and wet-screening concrete by DT method; prediction for tensile softening curve of dam concrete over neural network method; size effect of fracture energy and softening curve determined by the inverse analysis method; effect of aggregate size on softening curve; determination of dam concrete's fracture energy and comparison of fracture energy between the fracture work method and softening curve method; effect of tested curve tail section processing on fracture energy of dam concrete; experimental research on double-K facture parameters of the Three Gorges dam concrete; double-K facture parameters of dam concrete with various graded aggregates; analysis and research on crack prevention in hole roof for the Three Gorges project; construction technology and practice of no cracks concrete dam.

During this research and while writing this book, various people including our collaborative research advisor, academician Guofan Zhao of Dalian University of Technology supported and provided the funds for working in Shandong Province to purchase advanced equipment DH5937 to conduct fracture experiments. The academicians Chaoran Zhang of China Three Gorges Corporation, Chuhan Zhang of Tsinghua University and other experts have personally visited the site several times to inspect and direct these experiments. Their kind assistance greatly improvd the experiments and quality of the research. Additionally, Yangtze River Scholars, Professor Qingbin Li of Tsinghua University, Professor Shiliang Xu of Dalian University of Technology, Professor Kejin Wang of Iowa State University, Professor Volker Slowik of Hochschule für Technik, Wirtschaft und Kultur Leipzig, Dr. Seung Hee Kown of Northwestern University and Professor Jun Zhang of Tsinghua University provided tremendous help and direction for this research. The experimental research received assistance from Duanming Wang, Jingang Ma, Kaiyan Tan, Yongjun Song, Ke Zhang, Shouyang Zhao and Yanlin Zhang of CGGC, as well as Fude Zhang and Wencui Zhang of Tsinghua University. Dr. Zhigang Zhao, a teacher of Zhejiang University of Media and Communications has devoted himself to develop the necessary computer program since 2002, thus making great contributions to this research and was awarded not only the Chinese national software copyright, but the Science and Technology Achievement Award from Zhejiang Province. Our American friend, Henry Landan helped us to refine the book in English. The postgraduates Bin Zhang, Dongming Zhu, Ruixin Jin, Jiamin Dai, Jianping Chen, and Jintao Cai of Zhejiang University of Technology helped to draw some figures for the book, and we sincerely appreciate their valuable contributions.

Finally, We especially extend our profound respect and thanks to professors Chaoran Zhang and S. P. Shah for taking time from their busy schedules to write the Prologues for this book.

Zhejiang University of Technology devoted large amount of manpower, materials and funds to the research and publication of this book. This book was funded by the Monograph and Postgraduate Textbook Publication Fund of Zhejiang University of Technology. It is highly appreciated.

The research was funded by the Natural Science Foundation of China (51479178, 50409005), Natural Science Foundation of Zhejiang Province (LY14E090006, Y1100757, Y106486), Natural Science Foundation of Shandong Province (Q2001F02), Project (GDDCE15-01, GDDCE14-01) supported by Guangdong Provincial Key Laboratory of Durability for Civil Engineering (Shenzhen University), and Postgraduate Education Reform Project (2016-ZX-236).

The authors also wish to state that readers' suggestions and comments are encouraged and we warmly welcomed.

Authors
June, 2017

Contents

序 I ·· 3

Preface I ··· 5

Preface II ·· 7

Summary ··· 9

1 Introduction ··· 1
    1.1 Overview ··· 1
    1.2 Fracture Experiments of Three Gorges Dam Concrete ···································· 2
        1.2.1 Specimens ··· 2
        1.2.2 Materials and mix proportions ·· 3
        1.2.3 Test program ··· 3
        1.2.4 Three-point bending notched beam (TPB) tests ······································ 5
        1.2.5 Wedge splitting (WS) test ··· 8
        1.2.6 Direct tension (DT) tests and fundamental mechanical performance tests
        ··· 12

2 State-of-the-art Review on Concrete Softening Curve ··· 14
    2.1 Introduction ·· 14
    2.2 Determination Approach of the Tensile Softening Relationships (σ-w curves)
        of Concrete ··· 14
        2.2.1 Direct tension test method ··· 14
        2.2.2 J-Integral method ··· 15
        2.2.3 Inverse analysis method ··· 16
    2.3 Shape of Softening Curve of Concrete ·· 16
        2.3.1 Linear shape softening curve of concrete ··· 16
        2.3.2 Linear softening curve of concrete ·· 17
    2.4 Conclusions ·· 19
    References ··· 19

3 Two Methods for Determining Softening Relationships of Dam Concrete and Wet-screening Concrete ··· 21
    3.1 Introduction ·· 21

  3.2 Experiments ·· 22
    3.2.1 Material, mix and specimens preparation ························· 22
    3.2.2 Fracture tests ·· 23
  3.3 Softening Relationships of Dam Concrete and Wet-screened Concrete Determined by the Direct Tension Tests ·· 25
    3.3.1 Stress-deformation (σ-δ) curves ··· 25
    3.3.2 Direct tension method for identifying σ-w curve ··············· 25
    3.3.3 Results of direct tension method ····································· 26
  3.4 Softening Relationships Determined by the Inverse Analysis Method ············ 28
    3.4.1 Cracking strength ··· 28
    3.4.2 Inverse analysis method based on the cracking strength criterion ········· 28
    3.4.3 Results of inverse analysis method ································· 31
  3.5 Comparison of Softening Relationships of Dam Concrete and Wet-screening Concrete ·· 33
    3.5.1 Comparison of Softening Relationship of Dam Concrete and Wet-Screening Concrete Determined by Direct Tension Method and Inverse Analysis Method ··· 33
    3.5.2 Comparison of Fracture Energy of Dam Concrete and Wet-Screening Concrete Determined by WOF Method and TSD Method ·················· 35
  3.6 Conclusions ·· 36
  References ··· 37

4 Prediction of the Tension Softening Curve of Dam Concrete Based on BP Neural Network ·· 39
  4.1 Introduction ·· 39
  4.2 Introduction of the Direct Tension Test ··· 39
  4.3 Prediction of the Tension Softening Curve of Dam Concrete ··············· 39
    4.3.1 Determination of the network structure ··························· 40
    4.3.2 Selection of network transfer function ···························· 40
    4.3.3 Training and simulation of the network ·························· 41
  4.4 Conclusions ·· 44
  References ··· 44

5 Effect of Specimen Size on Fracture Energy and Softening Curve of Concrete: Part I. Experiments and Fracture Energy ··· 46
  5.1 Introduction ·· 46
  5.2 Experiments ·· 46
    5.2.1 Materials ·· 46
    5.2.2 Test program ·· 47

	5.2.3	Specimen and test set-up ·· 48
	5.3	Data Processing for Companion Specimens ··· 51
	5.3.1	Averaging the data for companion specimens ································ 51
	5.3.2	Extracting the data points representing the load-CMOD curve ············ 52
	5.4	Test Results and Discussion ··· 54
	5.5	Conclusions ·· 62
	References ··· 63

6 Effect of Specimen Size on Fracture Energy and Softening Curve of Concrete: Part II. Inverse Analysis and Softening Curve ·· 65

    6.1 Introduction ·· 65
    6.2 Inverse Analysis and Softening Curve ··· 66
        6.2.1 General ·· 66
        6.2.2 Procedure of inverse analysis ··· 67
        6.2.3 Softening curves from the inverse analysis and averaging softening curves ··· 69
    6.3 Analysis Results and Discussion ·· 71
        6.3.1 Comparison between the measured and calculated peak load and CMOD at peak ·· 71
        6.3.2 Effect of ligament length on softening curve ································· 75
        6.3.3 Comparison between softening curves of beam and wedge splitting tests ··· 76
        6.3.4 Possible mechanism for the size and geometry effect of fracture energy ······· 76
    6.4 Conclusions ·· 78
    References ··· 79

7 Influence of Coarse Aggregate Size on Softening Curve of Concrete ···················· 81

    7.1 Introduction ·· 81
    7.2 Experiments ·· 81
    7.3 Inverse Analysis of Softening Curves of Concrete ··································· 82
    7.4 Procedure of Inverse Analysis ··· 82
    7.5 Determination of Initial Softening Parameters ······································· 83
    7.6 FEM Mesh Generation for the Inverse Analysis Calculation ····················· 84
    7.7 Results of Inverse Analysis Calculation ··· 85
    7.8 Influence of Maximum Aggregate Size on Softening Curve of Concrete ········ 86
    7.9 Conclusions ·· 87
    References ··· 87

8 Research on Softening Curve of Concrete and Fracture Energy by Different Methods ········ 88
  8.1 Introduction ········ 88
  8.2 Experiment and Data Processing ········ 89
    8.2.1 Experiment ········ 89
    8.2.2 Data Processing ········ 90
  8.3 Levenberg-Marquardt Optimization Algorithm ········ 90
  8.4 Inverse Analysis and Results ········ 91
  8.5 Comparison and Analysis of Concrete Fracture Energy by Two Methods ········ 93
    8.5.1 The 1st method recommended by RILEM to determine concrete fracture energy ········ 93
    8.5.2 The 2nd method based on softening curve to determine concrete fracture energy ········ 93
    8.5.3 Comparison of concrete fracture energy by the two methods ········ 93
  8.6 Conclusions ········ 94
  References ········ 94

9 Research on Softening Properties of Concrete Based on Three-point Bend Beam Tests ········ 96
  9.1 Introduction ········ 96
  9.2 Three-point Bend Notched Beam Tests ········ 96
    9.2.1 Materials and mixes ········ 96
    9.2.2 Test procedure ········ 98
  9.3 Softening Curve of Concrete Calculated by Inverse Analysis Method ········ 98
    9.3.1 Softening curve determined by FEM based on FCM model ········ 98
    9.3.2 Softening curve of concrete obtained by inverse analysis ········ 99
  9.4 Softening Behaviors of Concrete ········ 100
    9.4.1 Influence of coarse aggregate size of concrete on the softening curve ········ 101
    9.4.2 Influence of specimen size on softening curve of concrete ········ 102
    9.4.3 Influence of concrete strength on softening curve ········ 102
  9.5 Conclusions ········ 103
  References ········ 104

10 Softening Behaviors of Dam Concrete and Wet-screening Concrete ········ 105
  10.1 Introduction ········ 105
  10.2 Experiments ········ 105
    10.2.1 Mix propotions ········ 105
    10.2.2 Specimens ········ 107
    10.2.3 Direct tension method ········ 108

    10.2.4 Wedge splitting test method ·· 109
  10.3 Softening Curves of Dam Concrete and Wet-screening Concrete Determined by the Direct Tension Test Method ·· 110
  10.4 Experimental Results of Wedge Splitting Test and Test Data Processing for Inverse Analysis ·· 113
    10.4.1 Preliminary processing ·· 113
    10.4.2 Data smoothing ·· 114
  10.5 The Softening Curve of Dam and Wet-screening Concrete Determined by the Evolutionary Algorithm-based Inverse Analysis Method ·········· 116
    10.5.1 Original input data ·· 117
    10.5.2 Principle of FEM simulation of concrete crack propagation ·········· 118
    10.5.3 Mesh generation ·· 119
    10.5.4 Error function e ·· 119
    10.5.5 Optimization algorithm ·· 120
    10.5.6 Softening curves of dam concrete calculated by the inverse analysis method ·· 122
  10.6 Comparison between the Softening Curves by Inverse Analysis Method and Those by Direct Tension Method ·· 124
  10.7 Softening Behaviors of Dam Concrete and Wet-screening Concrete ·········· 126
    10.7.1 Effect of compressive strength on the softening curve ·········· 126
    10.7.2 Effect of aggregate size on the softening curve ·········· 127
    10.7.3 Effect of specimen size on the softening curve ·········· 128
    10.7.4 Comparison of softening curves between dam concrete and wet-screening concrete ·· 131
  10.8 Conclusions ·· 132
  References ·· 133

11 Effect of Processing of Tail Section of Tested Curve on Fracture Energy of Dam Concrete ·· 135

  11.1 Introduction ·· 135
  11.2 Determination of Fracture Energy of Dam Concrete and Wet-screening Concrete by Wedge-splitting Test ·· 135
    11.2.1 Processing of tail section of tested $P\text{-}CMOD$ curve for wedge-splitting specimen ·· 136
    11.2.2 Calculation of fracture energy of dam concrete and wet-screened concrete of wedge-splitting specimen ·· 138
    11.2.3 Comparative analysis of the fracture energy calculated by different methods ·· 139

  11.3 Discussion of the Fracture Energy of Dam Concrete and Wet-screened Concrete 140
  11.4 Conclusions 141
  References 141

12 Experimental Study for Determining Double-K Fracture Parameters of the Three Gorges Dam Concrete 143

  12.1 Introduction 143
  12.2 Double-K Fracture Parameters Based on the Cohesive Stress 143
    12.2.1 Critical effective crack length 144
    12.2.2 Cohesive toughness 144
    12.2.3 Unstable fracture toughness 145
    12.2.4 Fracture toughness 146
  12.3 Specimen Design and Test Methods 146
    12.3.1 Raw materials and mix propotion 146
    12.3.2 Specimen fabrication 146
    12.3.3 Test equipment 147
  12.4 Experimental Results and Analysis 147
    12.4.1 Uniaxial tension test 147
    12.4.2 TPB test 148
    12.4.3 Determining of double-K fracture parameters 148
  12.5 Conclusions 149
  References 150

13 Experimental Research on Double-K Fracture Parameters of Dam Concrete with Various Aggregate Gradation 151

  13.1 Introduction 151
  13.2 Experiments 152
    13.2.1 Concrete mix proportions and basic mechanical properties ... 152
    13.2.2 Specimens 153
    13.2.3 Arrangement of measuring points 153
  13.3 Determination of Initial Cracking Load P_{ini} 153
  13.4 Calculation Procedures and Results of the Double-K Fracture Parameters 159
    13.4.1 Calculation procedures 159
    13.4.2 Calculation of TPB specimen 159
    13.4.3 Calculation of WS specimen 160
    13.4.4 Experimental results 160

  13.5 Discussion on Wedge splitting Test Results ·· 160
  13.6 Size Effect and Geometry Effect of Fracture Toughness ····················· 172
    13.6.1 Size effect of fracture toughness ·· 172
    13.6.2 Geometry effect of fracture toughness·· 173
  13.7 Conclusions ·· 174
  References··· 175

14 Analysis and Research on Crack Prevention in Hole Roof for the Three Gorges Project ·· 176

  14.1 Presentation of the Problem ·· 176
  14.2 Causes and Analysis··· 176
    14.2.1 Paying no more attention to the construction procedure in designing ··· 177
    14.2.2 Ignorance of design intention in construction ························· 179
  14.3 Crack Control Measures ··· 181
  14.4 Conclusions ·· 182
  References··· 182

15 Construction Technology and Practice of Dam without Cracks ··················· 183

  15.1 Introduction ·· 183
  15.2 Dam Concrete Construction Technology ··· 183
    15.2.1 Concrete design: optimize the mix proportions of concrete ·········· 184
    15.2.2 Concrete mixing: strict control of exit temperature ················· 185
    15.2.3 Transportation of concrete: strengthening the insulation and cooling ······ 185
    15.2.4 Concrete pouring: continuous improvement of operation technology ······ 186
    15.2.5 Concrete care: fine implement of concrete curing ···················· 187
  15.3 Conclusions ·· 191
  References··· 191

1 Introduction

1.1 Overview

China Three Gorges Project (TGP), as the world's largest hydropower project, has its main benefits in flood control, power generation and navigation. The total amount of concrete necessary for the main works of the Three Gorges Project is well over 27 million cubic meters. It is an innovative work to study the durability and reliability of mass concrete by fracture mechanics.

The authors participated in the cooperative research Fracture Simulation Analysis of Mass Concrete borne by Tsinghua University, China Gezhouba (Group) Corporation (CGGC) and Yantai University in March, 2002 when the elevation 156m of construction of the TGP dam was reached. It took the authors more than 2 years to accomplish Fracture Experiments and Study of Three Gorges Project Dam Concrete which took the double-K fracture method as a starting point. The expenditure of these fracture experiments reached 918.6 thousand Yuan RMB at that time. This work is the largest scale and systematic experimental research on fracture characteristics of TGP dam concrete.

There are four categories specimens involved in the fracture experiments of TGP dam concrete, including direct tension (DT) specimens (11 series, 55 specimens) among which the largest specimen size is 250mm×250mm×500mm, three-point bending notched beam (TPB) specimens (20 series, 80 specimens) among which the largest specimen size is 2300mm×550mm×240mm, wedge splitting (WS) specimens (20 series, 80 specimens) among which the largest specimen size is 1200mm×1200mm×250mm and corresponding companion fundamental mechanical performance (FM) specimens. The specimens are composed of different specimen size, strength, aggregate size, geometry, dam and wet-screening concrete specimen series. There were 377 fracture specimens which the concrete volume reached 23.1 cubic meters. These specimens were prepared in construction site of TGP dam under the same condition as the preparation of TGP dam concrete.

The first stage research work has been completed by Feb. 2004 through achieving the double-K fracture parameters of TGP dam concrete. Subsequently the computer modeling of softening curve based on handy test for determining the softening relationship of dam concrete has been pursing. Thus, the softening curves and fracture energy were determined by these works.

The fracture parameters, fracture mechanism and computer modeling on dam concrete and wet-screening concrete DT, TPB and WS specimens were investigated systematically. The specimens include different mix proportions, specimen sizes and coarse aggregate sizes. Some technical innovations and important laws based on the experimental study were achieved.

The DT, TPB and WS tests of 11 different categories dam concrete were conducted by employing the advanced testing techniques. The loading device of fracture test for large size aggregate concrete was developed which is characterized by large stiffness, large space and loading controlled by displacement. The initiation cracking load and crack propagation law were measured by employing resistance strain gauge method which is an innovation technology developed by our research team. Through the systematic fracture experimental studies of dam concrete and wet-screening concrete, WS test is regarded as an ideal testing method for the fracture parameters of large size aggregate dam concrete. Based on the principle of concrete fracture mechanics, the softening curves and fracture energy of dam concrete and wet-screening concrete were determined by direct tension method. The softening curves of dam concrete were predicted by using BP neural network method. The softening curves and fracture energy of dam concrete and wet-screening concrete were achieved by employing the inverse analysis method based on Levenberg-Marquardt optimization algorithm and evolutionary algorithm respectively and the corresponding software were developed. The influence factors, rules and mechanism of softening curves of dam concrete and wet-screening concrete were obtained. The comparative analysis of the softening curves of dam concrete obtained by direct tension method and inverse analysis method was conducted as well as the softening curves of wet-screening concrete. It is shown that the inverse analysis method based on the simple test is a handy and reliable method to determine the softening curve of dam concrete. The double-K fracture parameters of dam concrete and wet-screening concrete were achieved. The practical application of the research results in the Three Gorges Project, provide an important basis for the crack control technology of dam concrete.

However, the effects of temperature on fracture behaviors of dam concrete are not included as the limitation of research at that time. Now the authors are taking charge of a National Natural Science Foundation of China project Thermal Cracking Behavior of Ultra-high Volume Fly Ash Conventional Dam Concrete (HVFA) at Early Age Based on the Temperature-Stress Test (Grant No. 51479178). In this project the physical and mechanical behaviors of dam concrete at early age were studied considering the effect of temperature by employing temperature matched curing. In the future the effect of temperature on fracture behaviors of mass concrete will be focused on and the related research results will be reported.

1.2 Fracture Experiments of Three Gorges Dam Concrete

1.2.1 Specimens

There were four categories specimens involved in the fracture experiments of TGP dam concrete, including DT specimens (11 series, 55 specimens), TPB specimens (20 series, 80 specimens), WS specimens (20 series, 80 specimens) and corresponding companion fundamental mechanical performance (FM) specimens. The specimens are composed of different specimen size, strength, coarse aggregate size, geometry, dam and wet-screening concrete specimen series.

The maximum aggregate size of dam concrete is 80mm. The actual graded concrete specimens were prepared as well as their two graded wet-screening concrete specimens. The wet-screening procedure is to remove all aggregates which size is larger than 40mm from fresh dam concrete. Wet-screenedning concrete specimens are widely adopted to conduct experiments, and it is assumed that the performance indices of wet-screening concrete are the same as those of the actual graded concrete. This is an approximate method. It is important to study the relationships of fracture parameters between the actual graded concrete (large aggregate size concrete) and the wet-screening concrete.

1.2.2 Materials and mix proportions

The DT, TPB, WS and FM specimens were prepared by 11 categories concrete. The concrete mix proportions for these specimens are listed in Table 1-1.

Type IV cement, fly ash, crushed gravel and sand were employed in all mixes. The 80mm, 40mm, 20mm and 10mm crushed gravels were employed as coarse aggregate for the mixes. The water-reducing agent and air-entraining agent were employed to improve the workability, durability and consolidation of concrete.

Mix proportions Table 1-1

Specimen designation		Design strength grade	Maximum aggregate size (mm)	Water /binder ratio	Unit weight (kg/m^3)						
					Cement	Fly ash	Water	Sand	Coarse aggregate	Water-reducing agent	Air-entraining agent
SG	1	C20	10	0.50	196	84	140	869	1090	1.680	0.0196
	2	C20	20	0.50	185	79	132	846	1152	1.584	0.0185
	3	C30	20	0.45	240	60	135	814	1154	1.800	0.0195
	4	C40	20	0.35	309	77	135	744	1145	2.316	0.0251
	5	C50	20	0.30	420	47	140	698	1121	2.802	0.0280
	6	C20	40	0.50	168	72	120	769	1287	1.440	0.0168
LG	1	C20	80	0.50	143	61	102	653	1491	1.224	0.0143
	2	C25	80	0.45	159	68	102	625	1496	1.362	0.0159
WG	1	C20	80	0.50	143	61	102	653	1491	1.224	0.0143
	2	C25	80	0.45	159	68	102	625	1496	1.362	0.0159

1.2.3 Test program

The fracture experiments including TPB tests and WS tests were conducted on a stiff testing machine in a deformation control mode in Testing Center of China Gezhouba (Group) Corporation. Simultaneously the fundamental mechanical characteristics of companion specimens such as f_{cc}, f_{ts}, E_c and μ were tested. The deformation rate of loading kept constant during the whole loading process. Table 1-2 shows the testing indexes and results.

Testing indexes and results　　　　　　　　　　　　　　　　Table 1-2

Test items	Specimen categories					Test results
	SL	WS	DT	FM-1	FM-2	
f_{cc}				★		f_{cc}
E_c					★	E_c
μ					★	μ
f_{ts}				★		f_{ts}
P-CMOD	★	★				P-CMOD
P-CTOD	★	★				P-CTOD
P-δ	★					P-δ
P_{ini}	★	★				P_{ini}
P-ε	★	★				P-Δa
f_t			★			f_t
E_t			★			E_t
P-ε			★			σ-ε, σ-δ, σ-w

　　SL: three-point bending notched beam specimen.
　　WS: wedge splitting specimen. DT: direct tension specimen.
　　FM-1: fundamental mechanical performance specimen, 150mm×150mm×150mm cube for testing the cubic compressive strength f_{cc} and the splitting tensile strength f_{ts}.
　　FM-2: fundamental mechanical performance specimen, $\phi 150 \times 300 mm^3$ cylinders for testing the static compressive elastic modulus E_c and the Possion's ratio μ.
　　f_{cc}: cubic compressive strength.
　　E_c: static compressive elastic modulus.
　　μ: Possion's ratio.
　　f_{ts}: splitting tensile strength.
　　P-CMOD: load-crack mouth opening displacement curve.
　　P-CTOD: load-crack tip opening displacement curve.
　　P-δ : load-deflection curve.
　　P_{ini}: initial cracking load.
　　P-ε : load-strain curve which were tested by arranging testing points along the crack propagation path for TPB or WS specimen.
　　P-Δa: load-crack extension length curve.
　　f_t: uniaxial tensile strength.
　　E_t: tensile elastic modulus.

P-ε: uniaxial tensile load-strain curve.

σ-ε: uniaxial tensile stress-strain curve.

σ-δ: uniaxial tensile stress-deformation curve.

σ-w: stress-crack opening width curve.

1.2.4 Three-point bending notched beam (TPB) tests

1.2.4.1 Geometry, dimension and quantity of the TPB specimens

The geometry of specimens is shown in Fig. 1-1, dimension and quantity are shown in Table 1-3. For each group there are 4 specimens, and testing 3 specimens successfully is eligible.

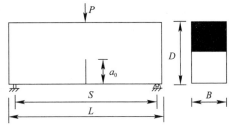

Fig. 1-1 Geometry of TPB specimen

TPB specimens   Table 1-3

Category		Specimen	Design strength grade	d_{max} (mm)	Dimensions (mm)			
					B	D	L	a_0
TPB specimen (20 groups, 80 specimens)	Small aggregate size concrete (48 specimens)	SL1 (SG1-B1)	C20	10		300	1300	120
		SL1y (SG1-B2)				400	1700	160
		SL2 (SG2-B1)		20	120	100	500	40
		SL3 (SG2-B2)				150	700	60
		SL4 (SG2-B3)				200	900	80
		SL5 (SG2-B4)				300	1300	120
		SL6 (SG2-B5)				400	1700	160
		SL7 (SG2-B6)				500	2100	200
		SL8 (SG3-B1)	C30			300	1300	120
		SL9 (SG4-B1)	C40					
		SL10 (SG5-B1)	C50					
		SL11 (SG6-B1)	C20	40				
	Large aggregate size concrete (16 specimens)	SL43 (LG2-B1)	C25	80	240	400	1700	160
		SL44 (LG2-B2)				450	1900	180
		SL45 (LG2-B3)				500	2100	200
		SL46 (LG2-B4)				550	2300	220
	Wet-screened concrete (16 specimens)	SL47 (WG2-B1)		40	120	200	900	80
		SL48 (WG2-B2)				250	1100	100
		SL49 (WG2-B3)				300	1300	120
		SL50 (WG2-B4)				400	1700	160

1.2.4.2 Test content

The load-deflection of load point (P-δ) curve, load-crack mouth opening displacement (P-$CMOD$) curve, load-crack tip opening displacement (P-$CTOD$) curve and initial cracking load P_{ini} were tested for TPB specimens. The crack extension P-Δa curves were tested for specimens SL43~SL50.

1.2.4.3 Test apparatus

The test apparatus for TPB tests is illustrated in Fig. 1-2.

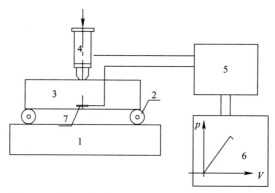

Fig. 1-2 Sketch of test apparatus for TPB specimen

1. Testing machine force bearing platform; 2. Roller shaft; 3. specimen; 4. load transducer; 5. Two DH5937 dynamic strain gauges; 6. Automatic acquisition system of computer; 7. displacement transducer

(1) Test equipments

The main test equipments are listed in the Table 1-4.

Test equipments Table 1-4

Order	Name	Type and specification	Manufacturer	Quantity
①	Compression testing machine	DY-500, Maximum load 5000kN	Changchun material testing machine factory	1
②	Controlling system of loading rate		Self-designed	1
③	Dynamic strain testing system	DH5937	Donghua Test Co.	2
④	Displacement sensor	CDP-5, Accuracy: 0.5μm, Range: 5mm	Japan	3
⑤	Load cell	BLR-1, Accuracy: 1%	Huadong electronic instrument factory	1
⑥	Resistance strain gauge	BE120-20AA Sensitive gird: 20.0mm×5.0mm	Zhongyuan electric measurement instrument factory	some
⑦	Data acquisition system	Matching with the dynamic strain testing system	Donghua Test Co.	1

(2) Instruments arrangement

The resistance strain gauges were arranged as shown in Fig. 1-3 for testing the crack extension: ① Distributed within the range of ligament and taking the pre-crack as symmetry axis. ② The first pairs were set at the pre-crack tip. Subsequently, there were 6 pairs which spacing was a_1, n pairs which spacing was a_2 and the last pairs which the surplus distance to the top of the beam was a_3. ③ The lateral spacing of the first pair of resistance strain gauges was $b_1=d_{max}$ (maximum aggregate size). b_2 rested upon the experiment.

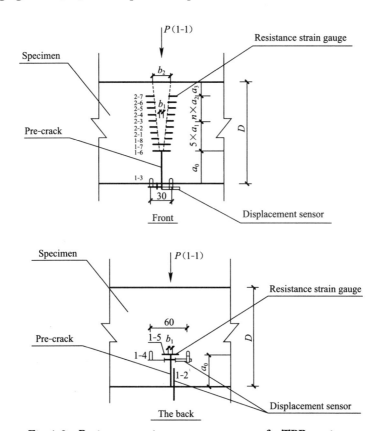

Fig. 1-3 Resistance strain gauges arrangement for TPB specimen

Parameters of arrangement of the resistance strain gauges on the façade of the beam   Table 1-5

D	a_1	a_2	n	a_3	Quantity of resistance strain gauges
200	20	0	0	20	12
250	20	20	2	10	16
300	20	20	3	20	18
400	20	30	4	20	20
450	25	30	4	25	20

continue

D	a_1	a_2	n	a_3	Quantity of resistance strain gauges
500	30	35	4	10	20
550	30	40	4	20	20

The channels of two dynamic strain testing systems are totally 16 channels. The testing item and requirement of each channel are depicted in Table 1-6.

Channels and testing items of the dynamic strain testing systems　　　Table 1-6

Channel	Instrument connected	Testing item	Requirement
1-1	Load cell	*Load-P*	Measuring accuracy ⩽2% maximum load
1-2	Displacement sensor	*Deflection-δ*	
1-3	Displacement sensor	*CMOD*	
1-4	Displacement sensor	*CTOD*	
1-5	Resistance strain gauge	P_{ini}	
1-6~2-8	Resistance strain gauge	Crack extension process	According to the sizes of the beams and aggregates, the resistance strain gauges connecting the channels were arranged (See Fig. 1-3)

1.2.5　Wedge splitting (WS) test

1.2.5.1　Geometry, dimension and quantity of the WS specimens

The geometry of WS specimen is shown in Fig. 1-4, dimension and quantity shown in Table 1-7. For each group there are 4 specimens, and testing 3 specimens successfully is eligible.

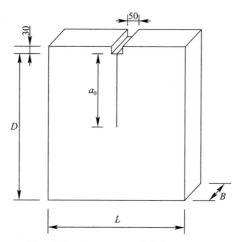

Fig. 1-4　Geometry of WS specimen

WS specimens Table 1-7

Category		Specimen	Design strength grade	d_{max} (mm)	Dimensions (mm)			
					B	D	L	a_0
WS specimen (20 groups, 80 specimens)	Small aggregate size concrete (44 specimens)	WS12 (SG1-W1)	C20	10	200	300	300	150
		WS12 (SG1-W2)				600	600	300
		WS13 (SG2-W1)		20		300	300	150
		WS14 (SG2-W2)				600	600	300
		WS15 (SG2-W3)				800	800	400
		WS16 (SG2-W4)				1000	1000	500
		WS17 (SG2-W5)				1200	1200	600
		WS18 (SG3-W1)	C30			300	300	150
		WS19 (SG4-W1)	C40					
		WS20 (SG5-W1)	C50					
		WS21 (SG6-W1)		40				
	Large aggregate size concrete (20 specimens)	WS22 (LG1-W1)	C20	80	250	450	450	225
		WS23 (LG1-W2)				600	600	300
		WS24 (LG1-W3)				800	800	400
		WS25 (LG1-W4)				1000	1000	500
		WS26 (LG1-W5)				1200	1200	600
	Wet-screening concrete (16 specimens)	WS32 (WG1-W1)	C20	40	200	300	300	150
		WS33 (WG1-W2)				600	600	300
		WS34 (WG1-W3)				800	800	400
		WS35 (WG1-W4)				1000	1000	500

1.2.5.2 Test content

The load-crack mouth opening displacement (P-$CMOD$) curve, load-crack tip opening displacement (P-$CTOD$) curve and initial cracking load P_{ini} were tested for WS specimens. The crack extension P-Δa curves were tested for specimens WS22-WS35.

1.2.5.3 Test apparatus

The test apparatus for WS tests is illustrated in Fig. 1-5.

1 Introduction

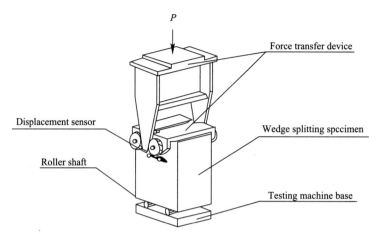

Fig. 1-5 Sketch of test apparatus for WS specimen

(1) Test equipments

The main test equipments are listed in the Table 1-8.

Test equipments and instruments　　　　　　　　　　Table 1-8

Order	Name	Type and specification	Manufacturer	Quantity
①	Compression testing machine	DY-500, Maximum load 5000kN	Changchun material testing machine factory	1
②	Controlling system of loading rate		Self-designed	1
③	Loading and force transfer equipment		Self-designed	1
④	dynamic strain testing system	DH5937	Donghua Test Co.	2
⑤	Displacement sensor	CDP-5, Accuracy: 0.5μm, Range: 5mm	Japan	4
⑥	Load cell	BLR-1, Accuracy: 1%	Huadong electronic instrument factory	1
⑦	Resistance strain gauge	BE120-20AA Sensitive gird: 20.0mm×5.0mm,	Zhongyuan electric measurement instrument factory	some
⑧	Data acquisition system	Matching with the dynamic strain testing system	Donghua Test Co.	1

(2) Instruments arrangement

The instruments arrangement is depicted in Fig. 1-6.

1.2 Fracture Experiments of Three Gorges Dam Concrete

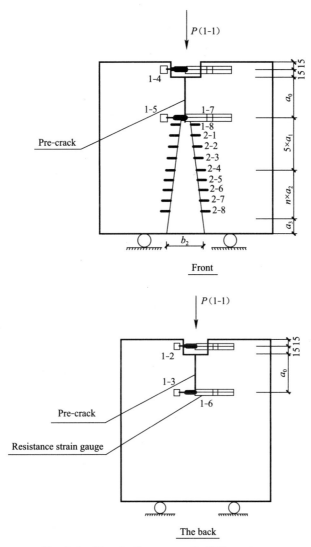

Fig. 1-6 Sketch of test setup for WS specimen

Parameters of arrangement of the resistance strain gauges on the facade of the specimen Table 1-9

D (mm)	a_1 (mm)	a_2 (mm)	n (piece)	a_3 (mm)	Quantity of resistance strain gauges (piece)
300	20	20	2	10	16
450	20	25	4	25	20
600	30	35	4	10	20
800	35	50	4	25	20
1000	45	65	4	15	20
1200	50	75	4	50	20

The channels of dynamic strain testing system are totally 16 channels. The testing item and requirement of each channel are depicted in Table 1-10.

1 Introduction

Channels and testing items of the dynamic strain testing systems　　Table 1-10

Channel	Instrument connected	Testing purpose	Requirement
1-1	Load cell	Load-P	Measuring accuracy ≤ 2% maximum load
1-2, 1-4	Displacement sensor	CMOD	
1-3, 1-5	Displacement sensor	CTOD	
1-6	Resistance strain gauge	P_{ini}	
1-7～2-8	Resistance strain gauge	Crack extension process	According to the sizes of the specimens and aggregates, the resistance strain gauges connecting the channels were arranged (See Fig. 1-6)

1.2.6　Direct tension (DT) tests and fundamental mechanical performance tests

1.2.6.1　The geometry, dimension and number of the specimens

Dimensions of specimens　　Table 1-11

Number	B (mm)	D (mm)	L Placement (mm)	L Finished (mm)	Quantity
b	ϕ 150		300	300	55
c	150	150	460	300	55
d	250	250	660	500	10

DT specimens　　Table 1-12

Category	Series	Batch	d_{max} (mm)	Design strength grade	Fracture specimens Groups SL	WS	WS$_S$	Subtotal	Direct tension and FM specimens Variety and quantity b	c	d	Subtotal
Small aggregate	Aggregate size	[1]	20	C20	6			6	5	5		10
		[2]	10		2	2		4	5	5		10
		[3]	40		1	1		2	5	5		10
	Strength	[4]	20	C20		5		5	5	5		10
		[5]		C30	1	1		2	5	5		10
		[6]		C40	1	1		2	5	5		10
		[7]		C50	1	1		2	5	5		10
Large aggregate		[8]	80	C20		4	2	6	5	5	5	15
		[9]				1	2	3	5	5		10
		[10]		C25		3	1	4	5	5	5	15
		[11]				2	1	3	5	5		10
Total					12	21	6	39	55	55		120

SL represents three-point bending notched beam, WS represents wedge splitting specimen, and WS$_s$ represents the wet-screening concrete for wedge splitting specimen. FM represents the fundamental mechanical specimens, including for testing compressive elastic modulus E_c and Possion's ratio μ.

Sizes of DT specimens Table 1-13

Specimen	Maximum aggregate size (mm)	Design strength grade	B (mm)	D (mm)	L (mm)
DT-1	20	20	150	150	300
DT-2	10	20	150	150	300
DT-3	40	20	150	150	300
DT-4	20	20	150	150	300
DT-5	20	30	150	150	300
DT-6	20	40	150	150	300
DT-7	20	50	150	150	300
DT-8	40	20	150	150	300
DT-9	80	20	250	250	500
DT-10	40	25	150	150	300
DT-11	80	25	250	250	500

1.2.6.2 Test content of DT test

The testing items were obtained by the DT test as follows: (1) load-strain curves in uniaxial tension. (2) stress-strain curves in uniaxial tension. (3) stress-deformation curves in uniaxial tension. (4) stress-crack width (σ-w) curves. (5) fracture energy G_F. (6) axial tensile strength f_t. (7) tensile elastic modulus E_t.

1.2.6.3 Fundamental mechanical performance test

The fundamental mechanical performance parameters were tested as follows: (1) cubic compressive strength (cube specimen: 150mm×150mm×150mm). (2) splitting tensile strength (cube specimen: 150mm×150mm×150mm). (3) static compressive elastic modulus E_c (Fig. 1-7 specimen b); (4) Possion's ratio μ (Fig. 1-7 specimen b).

The tests for the fundamental mechanical performance were conducted by employing a testing machine HSY1-015 with a load capacity of 2000kN in the Testing Center of CGGC. The tests were conducted simultaneously with the three types of fracture experiments above-mentioned.

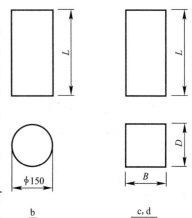

Fig. 1-7 Geometry of specimens

2 State-of-the-art Review on Concrete Softening Curve

2.1 Introduction

Since 1961, Kaplan[1] introduced fracture mechanics to the study of concrete materials, and researchers has been focusing on this field closely. Concrete is a composite with different material phases composed of coarse aggregate, fine aggregate and harden cementitious material. From the micro-level point of view, its remarkable characteristic is nonuniform, multi-phase and multi-porous. There also exists a large amount of micro-cracks between the aggregates and the cement matrix. Therefore, compared with metal materials, the fracture of concrete is even more complicated, which means there would be a micro-crack zone in front of a growing crack. Some experimental and numerical investigations has shown that the linear elastic fracture mechanics (LEFM) approach cannot applied to normal size concrete members. Hillerborg and his co-workers[2] made a major advance in concrete fracture by the fictitious crack model (FCM). No stress singularity of crack tip is present in the Hillerborg's model. Thereafter, some nonlinear fracture models were proposed to predict the fracture behavior of concrete such as crack band model (CBM)[3], two parameter fracture model (TPFM)[4], effective crack model (ECM)[5], size effect model (SEM)[6] and double-K fracture model[7], and so on.

In the fictitious crack model or cohesive crack model, the fracture characteristic of concrete is greatly influenced by the fracture process zone (FPZ). The inelastic fracture response due to FPZ may be considered by a cohesive stress acting on the crack faces. The cohesive stresses are modeled by an appropriate stress-crack opening σ-w relationship. The σ-w relation provides information on fracture resistance and could be employed for numerical simulations of crack propagation in concrete structures.

2.2 Determination Approach of the Tensile Softening Relationships (σ-w curves) of Concrete

The uniaxial tension test method, J-integral method and inverse analysis method are currently being employed to determine the tension softening curves of concrete.

2.2.1 Direct tension test method

Under ideal conditions, the softening curve could be directly achieved by direct tension tests of concrete. The fracture energy G_f corresponds to the area under the softening curve. However, the direct tension test has high requirements for the stiffness of a testing machine. Evans and Marathe[8] carried out the first tension test which showed post-peak response. They employed stiff bars paralleling to the specimen between the loading platens to take over the excess load. Even if the complete curve could be obtained by a material testing machine and a rigid frame, but there are still some indeterminable questions. Closed-loop testing has become available since 1970s. In a closed-loop test, the controlling factor is the elongation rather than the pressure in the actuator. To achieve complete stress-deformation curves in the direct tension (DT) test requires closed-loop control mode[9]. Specific problems in a DT test are the alignment of the specimen in the "loading chain" and the gripping of the specimen. The alignment will always remain difficult. The eccentricity is not always avoided. The question is to what extent eccentricities may be allowed before they will seriously affect the test results. In general, the tensile load will decrease with increasing load eccentricity[10] and the effect will be less influential when the heterogeneity of the material increases.

2.2.2 J-Integral method

Rice[11] proposed J-integral for characterizing the fracture toughness of metals. Li[12] proposed a J-integral method to experimentally identify the tension-softening (σ-w) relations for cementitious composites. The cohesive stress σ is a function of the crack opening width w. The J-integral can be expressed as the Eq. (2-1):

$$J = \int_0^w \sigma(w) dw \qquad (2\text{-}1)$$

A critical value of $J = J_c$ is reached when w at the physical crack tip reaches w_c. In this case J_c may be interpreted as the complete area under the tension-softening curve and w_c is the maximum crack width at which the load carrying capacity just vanishes.

The tension-softening relation can be got by differentiation as the Eq. (2-2):

$$\sigma(w) = \frac{\partial J(w)}{\partial w} \qquad (2\text{-}2)$$

The two geometrically similar specimens with a slight difference in notch lengths are employed in the J-integral method. J is determined experimentally from the difference of the area of the recorded load-displacement (P-δ) curves obtained from the tested specimens. For the determination of the softening relationship, the P-δ curve for both specimens, must be achieved from the test.

The J-integral method has the advantages of requiring only small specimens, a simple stroke controlled loading machine and is relatively stable. However, the J-integral method has the disadvantages for concrete. When unloading occurs in the material, especially during fracture process zone growth, the requirement of the J-integral that the path of unloading follows the path of loading may not be satisfied. In addition, the main problem of this method lies in its sensitivity to the scatter in the test data due to the heterogeneities of the geometrically identical specimens.

2.2.3 Inverse analysis method

In recent years, researchers began to identify the complete softening curve of concrete materials by employing the inverse analysis method. This method is based on fracture experiments and numerical simulations by using non-linear fracture mechanics. There are two kinds of approaches: (1) those that define preliminarily a softening curve in terms of certain parameters, such as by assuming a bilinear shape (depending on 4 parameters), tri-linear shape (depending on 6 parameters), and tetra-linear shape (depending on 8 parameters), and then employ certain optimization technique to fit the parameters to the data[13-14]; (2) those that use points on the load-displacement (*P-CMOD*) curve to determine a corresponding set of points on the softening curve, Which is called multi-linear approach[15]. A multi-linear softening curve is assumed and the individual slopes are identified one after another by adjusting a corresponding increment of the calculated *P-CMOD* curve to the experimental one. In this way, the softening curve is formed step by step while the crack is propagating in the simulation. The multi-linear approximation of softening curve can give a high performance of the optimization procedure. In addition, it needs no initial assumptions about the shape of the softening curve. However, an assumption concerning the initial cohesive stress as a starting point for the multi-linear softening curve may lower the objectivity of inverse analysis results.

2.3 Shape of Softening Curve of Concrete

The structural behavior of concrete during cracking can be described by nonlinear fracture mechanics models such as the cohesive crack model. The crack propagation is represented in this approach by a fictitious crack, which consists of the actual stress-free crack plus an inelastic fracture process zone, in which the cohesive stresses are modeled using an appropriate stress-crack opening (σ-w) relationship. Since the actual softening curve of concrete is a smooth curve, the various simplified softening curve forms were proposed to analyze the fracture process of concrete by employing the cohesive crack model.

2.3.1 Linear shape softening curve of concrete

In numerical analysis of crack propagation of concrete, simplified softening curves such as linear shape softening curves are usually employed due to their simplicity. Currently, the most common linear softening curves mainly include single linear softening curve, bilinear softening curve, and trilinear softening curve, etc.

(1) Single linear softening curve

Hillerborg[2] assumed the softening relationship of concrete as a straight line relationship in his fictitious model, as shown in Fig. 2-1 (*a*). Although it isn't precise enough, it still can characterize some mechanical behaviors of concrete well.

(2) Bilinear softening curve

The softening function is approximated by a bilinear function. This function is completely characterized when the four parameters are determined, as shown in Fig. 2-1 (*b*). The bilinear curve was initially proposed by Petersson[16] and the four parameters are identified as shown in Eq. (2-3):

$$\sigma_1 = \frac{f_t}{3}; \quad w_1 = 0.8\frac{G_F}{f_t}; \quad w_c = 3.6\frac{G_F}{f_t} \qquad (2-3)$$

Where f_t—tensile strength; σ_1—stress value of the kink point; w_1—tensile strength

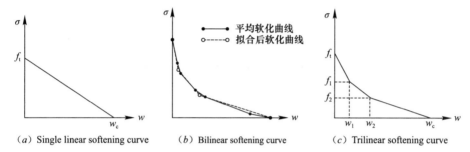

(*a*) Single linear softening curve (*b*) Bilinear softening curve (*c*) Trilinear softening curve

Fig. 2-1 Linear softening curves

The other curve[17] with some differences regarding the size of the tail and the location of the kink point in the bilinear softening curve was given by

$$\sigma_1 = \frac{f_t}{\alpha_F}\left(2 - \frac{f_t}{G_F}CTOD_c\right); \quad w_1 = CTOD_c; \quad w_c = \frac{\alpha_F}{f_t}G_F; \quad \alpha_F = \lambda - \frac{d_{max}}{8} \qquad (2-4)$$

Where $CTOD_c$—critical crack tip opening displacement; d_{max}—maximum aggregate size; λ—parameter related to the deformation behavior of concrete

(3) Tri-linear softening curve

The tri-linear softening curve, as shown in Fig. 2-1 (*c*), has been seldom mentioned in the literature. This could be because in most practical cases, a bilinear softening curve of concrete can give acceptable results, and also because it is harder to incorporate a tri-linear softening relationship in an analysis.

2.3.2 Linear softening curve of concrete

Some nonlinear softening models were also proposed, as shown in Fig. 2-2. Compared with the linear softening curves, nonlinear softening curve functions may have a better accuracy in the calculation.

1) Exponential softening curve

One of the simplest nonlinear softening curve functions is the exponential softening function. It was originally proposed by Gopalaratnam and Shah[18] in the following form:

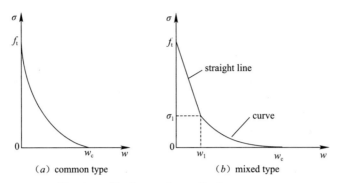

Fig. 2-2 Nonlinear shape of softening curve

$$\sigma = f_t \exp(-kw^\lambda) \quad (2\text{-}5)$$

Where, λ and k are constant values.

Li et al.[19] gave another exponential softening function in their research on the tensile behavior of high strength concrete.

$$\sigma = f_t \left\{ 1 - \phi \exp\left[-\left(\frac{\lambda}{w/w_c} \right)^n \right] \right\} \quad (2\text{-}6)$$

2) Power softening curve

Foote, Mai and Cotterell[20] proposed a power softening function to model crack propagation resistance of fiber cement composites. The function is expressed as:

$$\sigma = f_t \left(1 - \frac{w}{w_c} \right)^n \quad (2\text{-}7)$$

Where n is a fitting parameter and equal to 3.

3) Power-exponential softening curve

Reinhardt et al.[21] proposed a power-exponential softening function and given by:

$$\sigma = f_t \left[1 + \left(c_1 \frac{w}{w_c} \right)^3 \right] \exp\left(-c_2 \frac{w}{w_c} \right) - f_t \frac{w}{w_c} \left(1 + c_1^3 \right) \exp(-c_2) \quad (2\text{-}8)$$

Where c_1 and c_2 are constants which are determined from the test.

4) Mixed type softening curve

Jin et al.[22] proposed a mixed type softening curve by integrating the linear softening curve with the nonlinear softening curve to solve the following problem, as shown in Fig. 2-2 (b). Based on the analysis of bilinear softening curve and nonlinear curve, it is found that the initial slope of the bilinear softening curve can better control the ascending part of the P-δ curve and the peak load in the inverse analysis, but its descending part is difficult to coincide with the test data. Even if the descending part of the P-δ curve calculated by a nonlinear softening curve function is very close to the experimental values, the obtained peak load is difficult to control.

2.4 Conclusions

The cohesive crack model is an effective tool to simulate the mode I fracture behavior of concrete at the FPZ in which the material obey the tension softening relationship.

Several representative determination methods of softening curves of concrete were discussed. Direct tension test method has strict requirements for testing machine and has to deal with the load eccentricity. Thus, it is difficult to succeed. In addition, the test is expensive and time consuming. For J-integral method, a great error will be induced between the two curves by any error of the measured curve, which can give rise to distortion of the calculated σ-w curve. The inverse analysis method may provide a better approach for identifying the σ-w curve of concrete. The various linear softening curves and nonlinear softening curves were discussed for the fracture analysis of concrete.

References

[1] M. F. Kaplan, Jr. American Concrete Institute, 1961, 58: 591-610.

[2] A. Hillerborg, M. Modeer and P. E. Petersson. Cement and Concrete Research, 1976, 6(6): 773-781.

[3] Z. P. Bazant and B. H. Oh. Materials and Structures, RILEM, 1983, 16: 155-177.

[4] Y. S. Jenq and S.P. Shah. J. Eng. Mech., ASCE, 1985, 111(10): 1227-1241.

[5] P. Nallathambi and B.L. Karihaloo. Mag. Concrete Res., 1986, 38 (135):67-76.

[6] Z. P. Bazant, J. K. Kim and P. A. Pfeiffer. Journal of Structural Engineering, ASCE, 1986, 112(2):289-307.

[7] S. Xu and H. W. Renihardt. International J. Fracture, 1999, 98(2):151-177.

[8] R.H. Evans and M.S. Marathe. Materials and Structures, 1968, 1(1): 61-64.

[9] Z. LI, S. M. Kulkarni and S.P. Shah. Experimental Mechanics, 1993, 33(3): 181-188.

[10] F. P. Zhou. Some aspects of tensile fracture behaviour and structural response of cementitious materials. Report TVBM-1008, Division of Building Materials, Lund Institute of Technology, 1988.

[11] J.R. Rice. J. Appl. Mech., 1968, 35(2): 379-386.

[12] V. C. Li, C.M. Chan and C. K. Y. Leung. Cement and concrete research, 1987, 17 (3): 441-452.

[13] P. E. Roelfstra and F. H. Wittmann, In. Fracture Toughness and Fracture Energy of Concrete, editor by F. H. Wittmann, Elsevier, Amsterdam, 1986, pp.163-175.

[14] S. H. Kwon, Z.F. Zhao and S.P. Shah. Cement and Concrete Research, 2008, 38(8-9): 1061-1069.

[15] Y. Kitsutaka. Journal of Engineering Mechanics, ASCE, 1997, 123(5): 444-450.

[16] P. E. Pertersson. Crack growth and development of fracture zones in plain concrete and similar materials. Technical Report TVBM-1006, Division of Building Materials-Lund Institute of Technology, Lund-Sweden, 1981.

[17] Z. F. Zhao, S. L. Xu. Journal of Tsinghua University, 2000, 40 (S1): 110-113.

[18] V. S. Gopalaratnam, S.P. Shah. ACI Journal, 1985, 82 (3): 310-323.

[19] Q. B. Li and F. Ansari. ACI Materials Journal, 2000, 97(1): 49-57.

[20] R. M. L. Foote, Y. W. Mai and B. Cotterell. Journal of Mechanics and Physics of Solids, 1986, 34 (6): 593-607.

[21] H. W. Reinhardt, H. A. W. Cornelissen and D. A. Hordijk. Journal of Structural Engineering, ASCE, 1986, 112 (11): 2462-2477.

[22] Nan-gou Jin, Xian-yu JIN, Jian Shen and Ye Tian. Journal of Zhejiang University (Engineering Science), 2009, 43 (4): 732-737.

3 Two Methods for Determining Softening Relationships of Dam Concrete and Wet-screening Concrete

3.1 Introduction

The fictitious crack model (Hillerborg *et al*, 1976) and crack band model (Bazant and Oh, 1983) are typical approaches to simulate a fracture process of cement-based materials in structures. If the fracture process of civil engineering or hydraulic engineering structures is simulated by employing the models, the experimentally determined fracture parameters, such as tensile strength f_t, fracture energy G_f and softening curve (σ-w curve) are needed. Usually there are two ways to obtain the softening curves. One of them is the direct tension test method (Petersson, 1981; Gopalaratnum and Shah, 1985; Reinhardt *et al.*, 1986; Phillips and Zhang, 1993; Li and Ansari, 2000; Sousa and Gettu, 2006). This method has strict requirements for testing machine and has to deal with the loading concentricity which is often difficult to achieve. In addition, the test is expensive and time consuming. Inverse analysis is the other method (Sousa and Gettu, 2006; Planas *et al.*, 1999; Roelfstra and Wittmann, 1986; Kitsutaka *et al.*, 1997; Kitsutaka *et al.*, 2001; Zhang and Liu, 2003; Slowik *et al.*, 2006; Zhao *et al.*, 2008; Kwon *et al*, 2008; Zhang *et al.*, 2010). The softening curve can be achieved by a numerical procedure based on the three-point bending notched beam (TPB) test or wedge-splitting (WS) test which relatively easy to perform.

In This chapter, the direct tension (DT) tests and three-point bending notched beam (TPB) tests were performed on dam concrete and wet-screening concrete specimens. The specimens were made up of the dam concrete of which the mix was designed for an actual dam built in China. The maximum coarse aggregate size of dam concrete is 80mm. The wet-screening procedure is to remove all aggregate particles of which the sizes are larger than 40mm from the fresh dam concrete. Since it is difficult to test the large specimens made with 80mm maximum size aggregates, wet-screening concrete specimens are widely adopted to carry out experiments. It is assumed that the performance of wet-screening concrete may approximately characterize the dam concrete. This will be examined in this chapter. The softening relationships of dam concrete and wet-screening concrete were obtained by employing the above two methods respectively and the results are compared.

3.2 Experiments

3.2.1 Material, mix and specimens preparation

Type IV Portland cement was employed in the mix. The crushed gravels with maximum size of 80mm were used as coarse aggregate. The river sand was used as the fine aggregate in the mix. The water to binder ratio was 0.45 and the fly ash ratio was 30% of the total binder content. The water-reducing agent and air-entraining agent were employed to improve the workability, durability and consolidation of concrete. The mixture proportions of dam concrete are given in Table 3-1.

Mix proportions of dam concrete   Table 3-1

Maximum aggregate size(mm)	Unit weight (kg/m³)						
	Cement	Fly ash	Water	Sand	Coarse aggregate	Water-reducing agent	Air-entraining agent
80	159	68	102	625	1496	1.362	0.0159

The mix was designed for an actual dam built in China. Freshly cast specimens were kept in the moulds for 24 hours and subsequently they were demoulded and covered with the starw bags, cured by watering for 28 days and then kept in the same environmental condition (annual average temperature 18 celsius degree, annual average relative humidity 76%) as the dam for one year.

Five companion specimens and four companion specimens were manufactured for each DT test variable and each TPB test variable respectively. However, some specimens were broken in handing and moving, and reliable test results could not be achieved for some specimens. The numbers of companion specimens which were employed in DT and TPB tests and the reliable test data are listed in Table 3-2 and Table 3-3.

3.2.1.1 Preparation of the direct tension specimens

Sizes of the DT specimens are: 250mm×250mm×660mm for the dam concrete specimens, and 150mm×150mm×460mm for the wet-screening concrete specimens. Both sets were cast in the steel moulds. The length of 80mm was cut from both ends of the specimens by disc cutting machine before the tests. Then, the ends were polished by grinding machine. Shapes and sizes of the specimens are shown in Fig. 3-1 and Table 3-2 respectively.

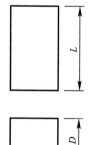

Fig. 3-1 Shape of DT specimen

3.2 Experiments

Sizes of DT test specimens Table 3-2

category	d_{max} (mm)	Specimens No.	Number of companion specimens	B (mm)	D (mm)	L (mm)
dam concrete	80	DT71~75	3	250	250	500
wet-screening concrete	40	DT66~70	3	150	150	300

d_{max} Maximum size of coarse aggregate.

3.2.1.2 Preparation of three-point bending notched beams

Sizes of the TPB specimens are: 1700mm×400mm×240mm for the dam concrete specimens, and 900mm×200mm×120mm for the wet-screening concrete specimens. The specimens were cast in wooden moulds and pre-formed crack was fabricated by the steel plate with thickness of 2mm, which was coated with lubricating oil on both sides. The steel plate was pull-out from concrete 3 hours before the concrete began to solidify. Sizes of the beams are shown in Table 3-3. In the table, B, D, L and a_0 are the width, depth, length and initial crack length of the TPB specimen, respectively.

Sizes of TPB specimens Table 3-3

Category	Specimen	Number of companion specimens	Span/depth ratio	Initial crack length/ depth ratio	Specimen sizes(mm)			
					B	D	L	a_0
Dam concrete	SL43	3	4	0.4	240	400	1700	160
Wet-screening concrete	SL47	3	4	0.4	120	200	900	80

3.2.2 Fracture tests

3.2.2.1 Direct tension test

(1) Test procedure

To obtain complete stress-deformation response in DT test, a very stiff testing machine and closed loop control machine is needed (Shah *et al.*, 1995). Also proper grips and extreme care in alignment of load line of action are important in order to avoid eccentric loading. The DT tests were conducted by employing a very stiff servo-hydraulic closed-loop testing machine with stiffness of 6000kN/mm and load capacity of 3000kN. The DT test is shown in Fig. 3-2.

Specimens were glued to the end platens of testing machine with epoxy. Four extensometers were installed around the specimen symmetrically for measuring the uniaxial tension deformation. The extensometers have a displacement range of ±5 mm. The average displacement from the four extensometers was used to calculate tensile strain. The maximum displacement from the output of four extensometers was employed as the feedback signal to control the machine. The speed used for strain control is 4-6με/min.

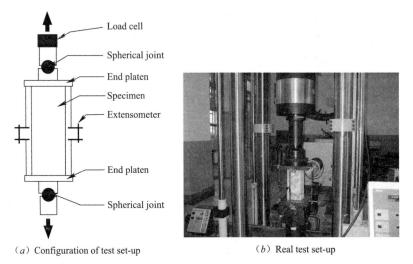

(a) Configuration of test set-up (b) Real test set-up

Fig. 3-2 Direct tension test

(2) Cracking and failure of DT specimens

The actual load-carrying capacity and stress distribution of each section of a specimen is different because the inhomogeneity which could result from the random distribution of coarse aggregates and the original flaws including initial cracks and voids. The cracks occurred first at the weakest section of the specimen. Note that eccentricity may not be avoided when a crack initiates from one side of the specimen. The issues of possible eccentricity and the end conditions have been dealt in details by others (Vervuurt et al., 1996, van Mier et al., 1996). The axial tensile failure sections of dam concrete specimens DT73、DT74 and DT75 were close to the middle part of the specimens as well as the wet-screening concrete specimen DT66. However, the failure section of wet-screening concrete specimens DT67 and DT69 located near the lower part of the specimens. The cracks of most specimens initiated from one side, then propagated toward the other side. However, the minority of specimens initiated from two sides, then propagated in ward.

3.2.2.2 Three-point bending notched beam test

(1) Test procedure

The tests were conducted on a very stiff testing machine. A load cell with a capacity of 100kN was adopted in the tests, and the accuracy was ±2% of the maximum applied load. The crack mouth opening displacement (CMOD) was measured by a displacement sensor whose capacity and accuracy are 5mm and 0.5μm respectively. It was possible to obtain the complete load-CMOD curves (see Fig. 3-3).

(2) Load-crack mouth opening displacement P-CMOD curves

The tested *P-CMOD* curves of dam concrete and wet-screening concrete are shown in Fig. 3-4. It is obvious that the maximum load p_{max} of dam concrete is higher than that of wet-screening concrete.

3.3 Softening Relationships of Dam Concrete and Wet-screened Concrete Determined by the Direct Tension Tests

(a) Dam concrete (b) Wet-screening concrete

Fig. 3-3 TPB tests of dam and wet-screening concrete

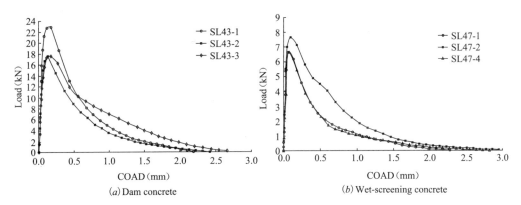

(a) Dam concrete (b) Wet-screening concrete

Fig. 3-4 Tested *P-CMOD* curves of dam concrete and wet-screening concrete by the three-point bend tests

3.3 Softening Relationships of Dam Concrete and Wet-screened Concrete Determined by the Direct Tension Tests

3.3.1 Stress-deformation (σ-δ) curves

The measured stress-deformation curves (σ-δ curves) of dam concrete and wet-screening concrete by the DT tests are shown in Fig. 3-5.

3.3.2 Direct tension method for identifying σ-w curve

From the σ-δ curves, σ-w curves were obtained by the following procedure and the sketch shown in Fig. 3-6.

The total deformation δ tested by the DT test can be expressed simply by superposition:

$$\delta = \delta_e + \delta_0 + w \tag{3-1}$$

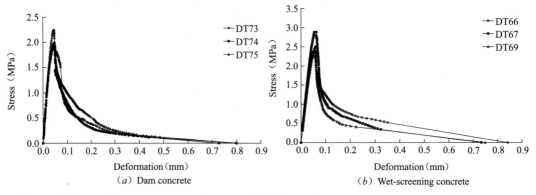

Fig. 3-5 Stress-deformation curves (σ-δ curves) of dam and wet-screening concrete in direct tension

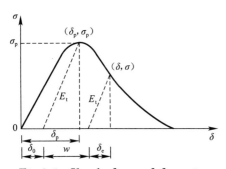

Fig. 3-6 Sketch of stress-deformation curve of concrete in tension

Where δ_e—elastic deformation at any point on the axial stress-deformation curve outside fracture zone;

δ_0—residual deformation at the peak stress point on the axial stress-deformation curve;

w—Crack width.

The elastic deformation δ_e can be calculated as follows:

$$\delta_e = \frac{\sigma}{E_t} l \quad (3\text{-}2)$$

Where E_t is modulus in tension;

l is length of specimen.

The residual deformation δ_0 can be determined by the following formula:

$$\delta_0 = \delta_p - \delta_{ep} \quad (3\text{-}3)$$

Where δ_{ep} is elastic deformation at the peak stress point on the axial stress-deformation curve; δ_p is deformation of specimen at peak stress.

So, the crack opening width w can be expressed as

$$w = \delta - \delta_e - \delta_0 \quad (3\text{-}4)$$

3.3.3 Results of direct tension method

The obtained σ-w curves are shown in Fig. 3-7. The relationship of relative uniaxial stress versus relative crack width (σ/f_t-w/w_0 curves) can be obtained through the normalization procedure and are shown in Fig. 3-8.

It can be seen that the tensile strength f_t and the fracture energy G_f of dam concrete are less than those of wet-screening concrete. The stress-free crack width w_0 of dam concrete is a

3.3 Softening Relationships of Dam Concrete and Wet-screened Concrete Determined by the Direct Tension Tests

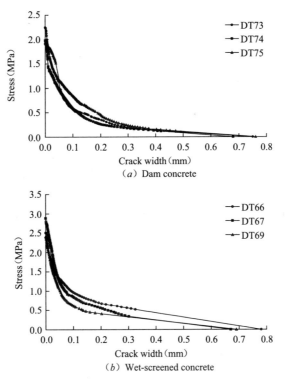

Fig. 3-7 Stress-crack width curves (σ-w curves) of dam and wet-screening concrete

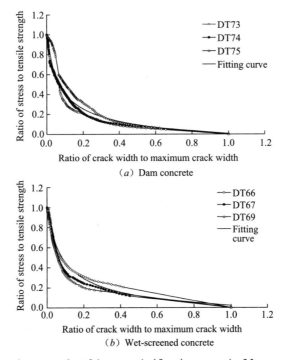

Fig. 3-8 Relative stress-relative crack width curves (σ/f_t-w/w_0 curves) of dam concrete and wet-screening concrete in direct tension

little larger than that of wet-screening concrete. w_0 is the maximum crack width at which the stress drops to zero. The expression of softening relationship can be obtained by a least-squares approach as follows:

$$\sigma = f_t \left\{ 1 - \varphi \exp\left[-\left(\frac{\lambda}{\dfrac{w}{w_0}} \right) \right]^n \right\} \tag{3-5}$$

in which φ、λ、n are material parameters. The specific parameters of the two kinds of concrete are given in Table 3-4. R^2 is least square correlative coefficient.

Softening relationships and fracture parameters determined by DT test   Table 3-4

Category	φ	λ	n	R_2	f_t (MPa)	w_0 (mm)	G_f (N/m)
Dam concrete	1.210	0.055	0.560	0.9652	2.052	0.73	232.778
Wet-screened concrete	1.230	0.068	0.550	0.9847	2.601	0.71	337.943

3.4 Softening Relationships Determined by the Inverse Analysis Method

Based on the cohesive crack model and cracking strength fracture criterion, the softening curve is achieved by a numerical procedure.

3.4.1 Cracking strength

Fig. 3-9 shows a typical graph illustrating the determination of cracking load from a tested *P-CMOD* curve of TPB specimen. The Cracking Strength σ_{fc}, defined as the stress level at which initial crack starts to propagate, is determined directly from the tested *P-CMOD* curves of three point bending notched beams. According to elastic theory, for a given initial crack, the *CMOD* and external load (*P*) obeys a linear relationship before the initiation of cracking. After initial cracking, the linear relationship between *P* and *CMOD* exists no longer. Thus the cracking load P_{fc}, can be determined by the point where the *P-CMOD* curve deviates from the initial linear portion, as shown in Fig. 3-9. This point is regarded as the transition point from the linear-elastic stage to the nonlinear-elastic stage, at which a fictitious crack starts to develop. Based on the P_{fc} value, the corresponding σ_{fc} can be calculated by a finite element analysis.

3.4.2 Inverse analysis method based on the cracking strength criterion

The direction of crack propagation is perpendicular to the maximum principle tensile stress in a TPB specimen shown in Fig. 3-10. Thereby, the path of crack propagation is parallel to the

3.4 Softening Relationships Determined by the Inverse Analysis Method

direction of load from the tip of pre-crack up to the top of the beam. Here the fictitious crack tip is defined as the point where the principle tensile stress attains the cracking strength σ_{fc} and the crack opening at this point is equal to zero. The stress can be transferred in the cohesive crack zone which is related to the crack width.

The softening curve (σ-w curve) can be simulated as piecewise linear function

$$\sigma_n = k_n w + \sigma_{0n},\ w_{n-1} \leqslant w \leqslant w_n, (n = 1, 2, \cdots, n_{max})$$

$$\sigma_{01} = \sigma_{fc},\ \sigma_{0n} = \sum_{1}^{n-1}[(k_i - k_{i+1})w_i] + \sigma_{fc} \qquad (3\text{-}6)$$

Where k_n and σ_{0n} are the slop and intercept of the nth line segment of the softening curve respectively, n_{max} is the number of the line segments.

The purpose of inverse analysis method is to determine the k_n of each segment from the tested P-CMOD curve, and then obtain the σ-w relation. A typical TPB specimen with pre-crack length a_0, crack length a, applied load P and cohesive stress $\sigma_b(w(x))$ acting on the crack surfaces is shown in Fig. 3-9. According to the principle of superposition illustrated in Fig. 3-9, the crack opening along the crack length can be expressed as follows (Balarin et al., 1994; Zhang et al., 2010)

$$w(x) = K_{pw}(x)P - B\int_0^a K_{fw}(x, y)\sigma_b(y)dy \quad 0 \leqslant x < a \qquad (3\text{-}7)$$

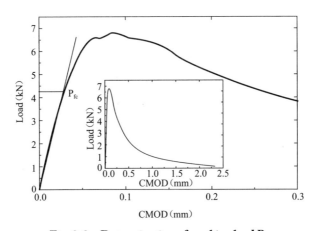

Fig. 3-9 Determination of cracking load P_{fc}

Similarly, the crack tip stress can be got by:

$$\sigma_{tip} = K_{p\sigma}P - B\int_0^a K_{f\sigma}(x, y)\sigma_b(y)dy \quad 0 \leqslant x < a \qquad (3\text{-}8)$$

When the equilibrium is reached, $\sigma_{tip} = \sigma_{fc}$. The crack mouth opening displacement CMOD (namely W_o) can be evaluated by the following Equation:

$$W_o = K_{pw-o}P - B\int_0^a K_{fw-o}(y)\sigma_b(y)dy \qquad (3\text{-}9)$$

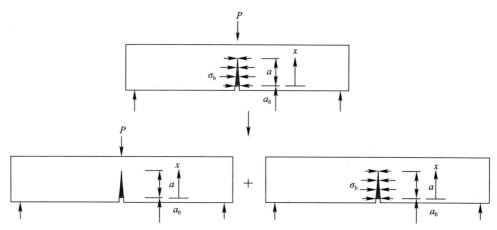

Fig. 3-10 Principle of superposition of cracking TPB under bending load and cohesive force

Where, K_{pw}, K_{fw} and $K_{p\sigma}$, $K_{f\sigma}$ are the influencing factors of applied load and cohesive force on the crack opening and crack tip stress, respectively.

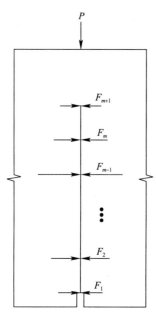

Fig. 3-11 Distribution of F.E. nodes along the path of crack propagation and cohesive force

The equations mentioned above by integral can be solved by discrete approach with matrices. According to the TPB beam in Fig. 3-11, the nodes are arranged along the pre-setting fracture line. The cohesive force acting on the crack surface is replaced by the node force which follows σ-w relationship. According to the cracking strength fracture criterion, the node is separated two nodes on which a couple of opposite nodal force begins to act when the crack tip stress reaches the cracking strength σ_{fc}. Consequently the node of crack tip moves to the next node. If we set the vectors $w = (w_1, w_2, w_3, \cdots, w_m)$, $F = (F_1, F_2, F_3, \cdots, F_m)$ and assume the node number of crack tip being $(m+1)$, then the equations (3-7) and (3-8) can be expressed by matrices as follows:

$$w_i = K_{pw-i}P - \sum_{i=1}^{m} K_{fw-i} F_i \quad (i=1,2,\cdots,m) \tag{3-10}$$

$$\sigma_{fc} = K_{p\sigma}P - \sum_{i=1}^{m} K_{f\sigma-i} F_i \quad (i=1,2,\cdots,m) \tag{3-11}$$

F_i is related to w_i by

$$F_i = \left(k_n w_i + \sigma_{0i}\right) B \Delta l_i \quad (i=1,2,\cdots,m) \tag{3-12}$$

Where B is the width of specimen, Δl_i is the calculating length at node i. $\Delta l_i = \Delta l$ except for the case when $i=1$, $\Delta l_i = 0.5\Delta l$. Here, Δl is the distance between two adjacent nodes.

The influence factors of applied load and cohesive force can be obtained by a finite element analysis. Thus the $2m+1$ equations including $2m+1$ unknowns are composed of the equations

3.4 Softening Relationships Determined by the Inverse Analysis Method

(3-10), (3-11) and (3-12) for the given length of crack $a(a=m\Delta l)$. Namely, the desired parameters can be obtained by solving the linear equation group containing the unknowns $x=(P, w_1, w_2, w_3, \cdots, w_m, F_1, F_2, F_3, \cdots, F_m)$. Meanwhile $CMOD$ can be evaluated by

$$w_o = K_{pw-o}P - \sum_{i=1}^{m} K_{fw-o-i} f_i \quad (i=1, 2, \cdots, m) \qquad (3\text{-}13)$$

The applied load P, cohesive force F and crack opening $w(x)$ can be obtained by solving the mentioned above equations for a given crack length a and σ-w relationship.

Employing crack length a as a variable, the loading process is simulated by numerical procedure according to the mentioned above model and equations. The least crack length is Δl and the crack propagates with the step increment Δl or multiple of Δl until the maximum length of crack $m_{max}\Delta l$. The difference between the calculated load and experimental load is made to be minimum by optimizing k in equation (3-6) for each step of crack propagation. Finally, the complete σ-w curve can be achieved. The evaluation details are presented in literature (Zhang and Liu, 2003). If a σ-w relationship is given, the fracture energy of material can be evaluated by:

$$G_f = \int_0^{w_c} \sigma(w) dw = \sum_{i=1}^{m_{max}} \sigma_i w_i \quad (i = 1, 2, \cdots, m_{max}) \qquad (3\text{-}14)$$

3.4.3 Results of inverse analysis method

Based on the tested P-$CMOD$ curves, the σ-w curves of dam concrete and wet-screening concrete were obtained by inverse analysis method, as shown in Fig. 3-12 (*a*) and Fig. 3-13 (*a*). In order to distinguish between the cohesive stress versus crack opening width relation (σ-w relation) based on the Hillerborg's fictitious crack concept and that based on the cracking strength concept mentioned in This chapter, σ-w is employed to express the softening relationship obtained by the inverse analysis based on the cracking strength concept in the Fig. 3-12 (*a*) and Fig. 3-13 (*a*). The Fig. 3-12 (*b*) and Fig. 3-13 (*b*) illustrates the comparison of calculated P-$CMOD$ curve with the tested one.

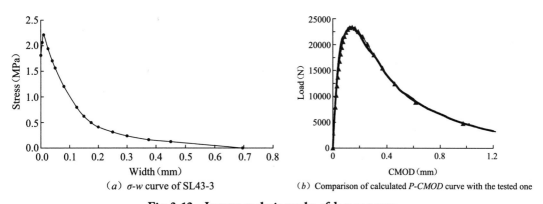

(*a*) σ-w curve of SL43-3   (*b*) Comparison of calculated *P-CMOD* curve with the tested one

Fig. 3-12 Inverse analysis results of dam concrete

3 Two Methods for Determining Softening Relationships of Dam Concrete and Wet-screening Concrete

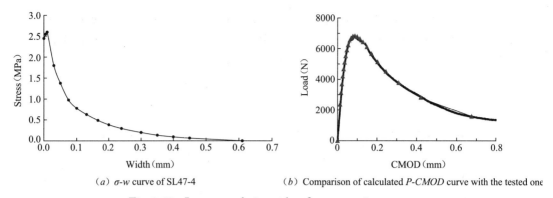

(a) σ-w curve of SL47-4   (b) Comparison of calculated P-$CMOD$ curve with the tested one

Fig. 3-13 Inverse analysis results of wet-screening concrete

As the ascending part of softening curve from cracking strength σ_{fc} to tensile strength f_t is approximately depicted by a part of stress-strain curve in the fictitious crack model (Hillerborg et al., 1976; Shah et al., 1995), so the softening curve starts from the point of which the ordinate equals to the tensile strength f_t. In order to compare the σ-w curves by the two methods, the mentioned above processing was employed to the σ-w curves obtained by the inverse analysis. Namely, the ascending part of softening curve from σ_{fc} to f_t is still depicted by the stress-strain curve and the original point of σ-w curve is set to the point of which the ordinate equals to f_t. According to the above principle, σ-w curve was determined for each specimen. The softening curve for each group of specimens can be achieved by fitting the data points of softening curves of the group composed of three companion specimens. The σ-w curves and relative stress-crack width curves (σ/f_t-w/w_0 curves) of dam concrete and wet-screening concrete by inverse analysis method are shown in Fig. 3-14 and Fig. 3-15.

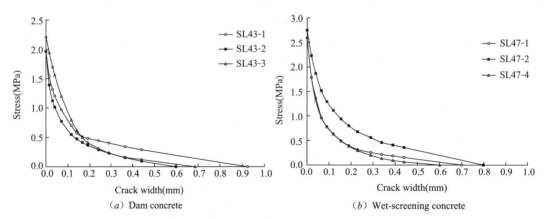

(a) Dam concrete   (b) Wet-screening concrete

Fig. 3-14 σ-w curves of dam concrete and wet-screening concrete by inverse analysis method

The softening relationships of dam concrete and wet-screening concrete through the least-squares approach are also given by the equation (3-5). The related parameters are listed in Table 3-5.

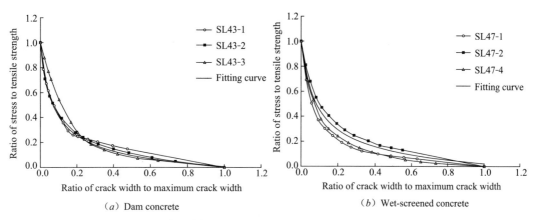

Fig. 3-15 σ/f_t-w/w_0 curves of dam and wet-screening concrete by inverse analysis method

Parameters of softening relationships and fracture parameters determined by inversed analysis method   Table 3-5

Category	φ	λ	n	R^2	f_t (MPa)	w_0 (mm)	G_f (N/m)
Dam concrete	1.210	0.055	0.560	0.9878	2.046	0.74	273.71
Wet-screening concrete	1.230	0.068	0.550	0.9879	2.645	0.70	321.58

3.5 Comparison of Softening Relationships of Dam Concrete and Wet-screening Concrete

3.5.1 Comparison of Softening Relationship of Dam Concrete and Wet-Screening Concrete Determined by Direct Tension Method and Inverse Analysis Method

It can be seen from Table 3-4 to Table 3-8 that the expressions of softening relationships obtained by the two methods are almost identical. The correlation coefficients of the fitting curve expression R^2 are all above 0.95 and the difference between them is very small. From the σ-w relationships by the two methods, important parameters such as the tensile strength f_t, the stress-free crack width w_0 and the fracture energy G_f were obtained, as shown in Table 3-6 and Table 3-7. These parameters by the inverse analysis method have a good agreement with those by the direct tension methods. The errors of the three parameters for dam concrete are 0.3%, 0.8% and 17.6%, and for wet-screening concrete are 1.7%, 1.5% and 4.8% respectively.

The comparison of the three important parameters of softening relationships for dam concrete and wet-screening concrete by the two methods is shown in Fig. 3-16. It shows that the f_t as well as the G_f of dam concrete were smaller than those of wet-screening concrete, but the w_0 of dam concrete was greater than that of wet-screening concrete. As the three parameters by the inverse analysis are concerned, the f_t, w_0 and G_f of wet-screening concrete are 1.29, 0.95 and 1.17 times the value of the dam concrete respectively. The softening curves for dam concrete and wet-screening concrete by the two methods can both be predicted by equation (3-5) and the related

3 Two Methods for Determining Softening Relationships of Dam Concrete and Wet-screening Concrete

parameters are listed in Table 3-4 and Table 3-5. The results show that there may be a definite relation between the softening relationship of dam concrete and that of wet-screening concrete.

Parameters of σ-w curves by the direct tension method   Table 3-6

Category	Specimen	f_t(MPa)	w_0 (mm)	G_f (N/m)
d_{max}=80mm dam concrete	DT73	2.248	0.75	225.803
	DT74	1.976	0.68	197.012
	DT75	1.931	0.76	275.519
	Average	2.052	0.73	232.78
d_{max}=40mm wet-screening concrete	DT66	2.508	0.78	421.362
	DT67	2.884	0.67	324.341
	DT69	2.411	0.69	268.127
	Average	2.601	0.71	337.94

Parameters of σ-w curves by the inverse analysis method   Table 3-7

Category	Specimen	f_t(MPa)	w_0 (mm)	G_f (N/m)
d_{max}=80mm dam concrete	SL43-1	1.963	0.924	338.65
	SL43-2	1.963	0.599	205.71
	SL43-3	2.213	0.686	276.77
	Average	2.046	0.736	273.71
d_{max}=40mm wet-screening concrete	SL47-1	2.588	0.698	262.86
	SL47-2	2.748	0.798	467.43
	SL47-4	2.598	0.600	234.46
	Average	2.645	0.699	321.58

Comparison of the parameters of σ-w curves between the two methods   Table 3-8

Category	Method	f_t(MPa)	w_0 (mm)	G_f (N/m)
d_{max}=80mm dam concrete	Inverse analysis method	2.046	0.736	273.71
	Direct tension method	2.052	0.73	232.78
	Error	-0.006	0.01	40.93
	Relative error	-0.003	0.008	0.176
d_{max}=40mm wet-screening concrete	Inverse analysis method	2.645	0.699	321.58
	Direct tension method	2.601	0.71	337.94
	Error	0.044	-0.01	-16.36
	Relative error	0.017	-0.015	-0.048

3.5 Comparison of Softening Relationships of Dam Concrete and Wet-screening Concrete

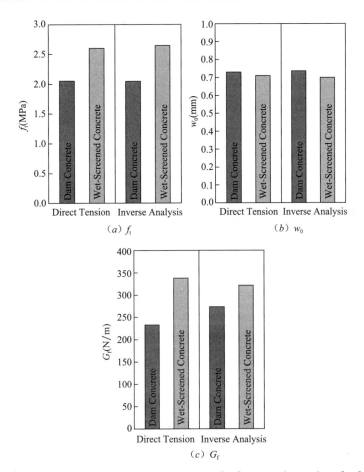

Fig. 3-16 Comparison of the three important parameters of softening relationships for dam concrete and wet-screening concrete by the two methods

3.5.2 Comparison of Fracture Energy of Dam Concrete and Wet-Screening Concrete Determined by WOF Method and TSD Method

The fracture energy of dam concrete and wet-screening concrete for TPB specimens were achieved based on RILEM recommendation (RILEM TC-50 FMC 1985), the results are listed in the Table 3-9. This method is called work of fracture method (WOF method). The other method which can achieve the fracture energy is tension-softening diagram (TSD) method. The results of direct tension tests were employed to achieve the fracture energy of dam concrete and wet-screening concrete by the TSD method, and the results are listed in the Table 3-9, too. The comparison of fracture energy of dam concrete and wet-screening concrete determined by WOF and TSD method is illustrated in the Fig. 3-17.

It was found that the self weight of TPB specimen has an effect on the fracture energy which was determined by WOF method. The effect of specimen self weight on fracture energy increases with increase of the specimen size. It is in accordance with the findings of the literature (Mindess 1984).

3 Two Methods for Determining Softening Relationships of Dam Concrete and Wet-screening Concrete

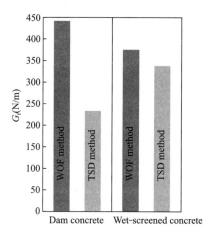

Fig. 3-17 Comparison of fracture energy of dam concrete and wet-screening concrete determined by WOF and TSD method

Fracture energy obtained by the TSD method and WOF method   Table 3-9

Category	Specimen	G_f by TSD method (N/m)	Specimen	G_f by WOF method (N/m)
d_{max}=80mm dam concrete	DT73	225.803	SL43-1	451.462
	DT74	197.012	SL43-2	426.358
	DT75	275.519	SL43-3	450.665
	Average	232.78	Average	442.83
d_{max}=40mm wet-screening concrete	DT66	421.362	SL47-1	330.194
	DT67	324.341	SL47-2	465.028
	DT69	268.127	SL47-4	331.153
	Average	337.94	Average	375.46

3.6 Conclusions

The dam concrete which contain large size aggregates are widely applied in construction of hydraulic engineering, so it's very important to investigate fracture properties of dam concrete. The fracture properties of dam concrete can be predicted by wet-screening concrete tests which are easier to handle. In This chapter, the direct tension (DT) tests and three point bend (TPB) tests were conducted on dam concrete and wet-screening concrete specimens. The maximum specimen size is 250mm×250mm×500mm for DT specimen and 1700mm×400mm×240mm for TPB specimen. The following conclusions can be drawn from the current experiments and analysis:

(1) From the DT tests, it was observed that the specimens tended to subject the eccentric load after cracking. This is a drawback of identifying the σ-w curves of dam concrete and wet-

screening concrete from DT method.

(2) The σ-w curves and the expressions of dam concrete and wet-screening concrete were obtained by the DT test results and also by the inverse analysis. It was revealed that the σ-w curves by the DT test method were in good agreement with those by the inverse analysis method.

(3) It was also found that the tensile strength f_t as well as the fracture energy G_f of dam concrete were smaller than those of wet-screening concrete, but the stress-free crack width w_0 of dam concrete was greater than that of wet-screening concrete. It can be seen there may be a definite relation between the softening relationship of dam concrete and that of wet-screening concrete.

(4) The effect of self weight of TPB specimen on fracture energy which was achieved by WOF method tends to increase with increase of the specimen size.

(5) The inverse analysis method may be a desirable approach to find the σ-w relations of dam concrete and wet-screening concrete.

References

[1] Bazant, Z.P. and Oh, B.H.. Crack band theory for fracture of concrete. Materials and Structures. 16 (1983) 155-177.

[2] Balarin R., Shah, S.P., Keer, L.M.. Crack Growth in Cement based Composites. Engineering Fracture Mechanics. (1994) 43-443.

[3] Gopalaratnum, V.S., Shah, S.P. Softening response of plain concrete in direct tension. ACI Materials Journal. 82(3)(1985) 310-323.

[4] Hillerborg, A., Modeer, M. and Petersson, P.E.. Analysis of crack formation and crack growth in concrete by means of fracture mechanics and finite elements. Cement and Concrete Research. 6(6) (1976) 773-782.

[5] Kitsutaka, Y., Uchida, Y., Mihashi, H., Kaneko, Y., Nakamura, S. and Kurihara, N. Draft on the JCI standard test method for determining tension softening properties of concrete. In: R. de Borst, J. Mazars, G. Pijaudier-Cabot and J.G.M. van Mier, Editors, Fracture Mechanics of Concrete Structures, Proceedings of Framcos-4 vol. 1, Lisse, The Netherlands. A.A.Balkema Publishers. (2001) 371-376.

[6] Kwon, SeungHee, Zhao, Zhifang, Shah, Surendra P. Effect of specimen size on fracture energy and softening curve of concrete. Part II: Inverse analysis and softening curve. Cement and Concrete Research. 38(8-9) (2008) 1061-1069.

[7] LiQingbin, Ansari, F. High strength concrete in uniaxial tension. ACI Materials Journal. 97(1) (2000) 49-57.

[8] Petersson P. E.. Crack growth and development of fracture zones in plain concrete and similar materials. Report TVBM-1006, Division of Building Materials, Lund Institute of Technology. (1981).

[9] Phillips, D., Zhang, B. Direct tension tests on notched and unnotched plain concrete specimens. Magazine of Concrete Research. 145(162) (1993).25-32.

[10] Reinhardt, H.W., Cornelissen, A.W. and Hordijk, D.A.. Tensile tests and failure analysis of concrete.

Journal of Structure Engineering.112 (1986) 2462-2477.

[11] Roelfstra, P.E. and Wittmann, F.H. Numerical method to link strain softening with failure of concrete. In: F.H. Wittmann, Editor, Fracture Toughness and Fracture Energy of Concrete, Amsterdam, Elsevier. (1986) 163-175.

[12] Sousa, J. and Gettu, R. Determining the tensile stress-crack opening curve of concrete by inverse analysis. Journal of Engineering Mechanics. 132(2)(2006) 141-148.

[13] Slowik, V., Villmann, B., Bretschneider, N. and T. Villmann.. Computational aspects of inverse analyses for determining softening curves of concrete. Computational Methods in Applied Mechanics and Engineering. 195(2006) 7223-7236.

[14] Shah, S.P., Swartz, S. E. and Ouyang, C. Fracture mechanics of concrete. John Wiley &Sons, Inc., New York: 605 Third Avenue. (1995).

[15] vanMier, J.G.M., Schlangen E. and Vervuurt, A. Tensile Cracking in Concrete and Sandstone, Part 2: Effect of Boundary Rotations, Materials and Structures (RILEM). 29(186) (1996) 87-96.

[16] Vervuurt, A., Schlangen, E. and van Mier, J.G.M. Tensile Cracking in Concrete and Sandstone, Part 1: Basic Instruments, Materials and Structures (RILEM). 29(185) (1996) 9-18.

[17] Zhang J., Liu Q. Determination of concrete fracture parameters from a three-point bending test. Tsinghua Science and Technology. 8(6) (2003) 726-733.

[18] Zhao, Zhifang, Kwon, SeungHee, Shah, S.P. Effect of specimen size on fracture energy and softening curve of concrete. Part I: Experiments and fracture energy. Cement and Concrete Research. 38(8-9) (2008) 1049-1060.

4 Prediction of the Tension Softening Curve of Dam Concrete Based on BP Neural Network

4.1 Introduction

The design, construction and safety assessment of dam all require studying the fracture properties of dam concrete systematically and deeply. The tension softening curve is one of the important contents of the research on fracture properties of dam concrete. The direct tension test method, J-integral method and inverse analysis method are widely employed to get the softening curves of concrete at present. Those methods get softening curves by different tests. This paper is aimed at predicting the tension softening curves of dam concrete by the BP neural networks based on the test data of direct tension tests.[1]

4.2 Introduction of the Direct Tension Test

The direct tension tests of dam concrete were conducted by employing the digital servo-hydraulic closed loop testing machine Instron 8506 3000kN with stiffness of 6000kN/mm at the Tsinghua university. We have performed the direct tension tests of 7 groups dam concrete specimens and obtained their uniaxial tension stress-strain curves. All the 7 mixtures were designed for the Three Gorges Project dam concrete.[2] According to the principle of concrete fracture mechanics, the softening curves of dam concrete were achieved based on the direct tension test data by the direct tension test method. The tests are described in the reference[3] in detail and the view of the test is shown in Fig 4-1.

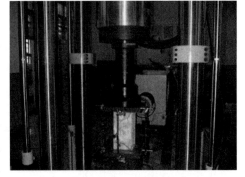

Fig 4-1 View of direct tension test of dam concrete

4.3 Prediction of the Tension Softening Curve of Dam Concrete

The cohesive crack model is widely used in the analysis of concrete crack extension. The softening relationship is an important constitutive relationship of this model. The Fig 4-2 shows

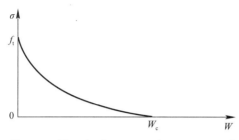

Fig 4-2 Sketch of tension softening curve of concrete

the sketch of softening curve of concrete. σ is cohesive stress, w is crack opening width, f_t is tensile strength of concrete and w_0 is maximum crack width.

Employing the BP neural network to establish training set based on the softening curve data of 6 groups dam concrete materials which were obtained by the above-mentioned direct tension test method, the softening curve of the 7th dam concrete material was predicted. The predicted softening curve of the 7th dam concrete material was compared with that obtained by the direct tension test method in order to verify the reliability of the prediction.

This paper mainly considers four factors water-binder ratio, sand ratio, maximum aggregate size and relative crack width (w/w_0) which have influence on softening curves and uses BP neural network to predict softening curves of dam concrete.

4.3.1 Determination of the network structure

The determination of BP neural network structure refers to determine the network layer number and the number of neuron units each layer. The BP network has very strong capability of nonlinear mapping. A three-layers BP neural network can realize the approximation to arbitrary nonlinear function (according to Kolrnogorov theorem).[4] The selection of the node number for each layer based on BP algorithm have an great influence on the network performance. The number of nodes inside layer should be selected appropriately. Four nodes of input layer were selected which are water-binder ratio (w/b), sand ratio (s/a), maximum aggregate size (Gmax/mm) and relative crack width (w/w_0). One node of output layer is relative stress (σ/f_t).

This research starts its pilot calculation from single hidden layer structure in determining the network structure. The study shows that the network can not have enough learning ability and capability of information processing if the number of hidden nodes is too less. If the number of hidden nodes is too numerous, it will not only increase the complexity of network structure greatly, but also make the network learning speed slow down.

In this paper, the selection of number of hidden layer nodes is determined preliminarily according to the proposal of Hecht-Nieisen who put forward the number of hidden layer nodes is 2N+1 (N is the number of input layer nodes)[5], and then it was adjusted in the learning process according to the error minimum principle . Therefore the optimal number of network hidden layer nodes is nine.

4.3.2 Selection of network transfer function

The transfer function represents the relationship between input and output of the processing unit.Neurons transfer function based on BP algorithm in the toolbox of MATLAB neural network includes linear function purelin, hyperbolic tangent s-type transfer function tansig and logarithmic s-type transfer function logsig.

When the transfer function is logsig between input layer and hidden layer, the same as that between hidden layer and output layer, the network structure is 4-9-1, and the operation result is as follows:

TRAINLM, Epoch 16/1000, MSE 0.00453963/0.005, Gradient 4.16732/1e-010

When the transfer function is tansig between input layer and hidden layer, the same as that between hidden layer and output layer, the other network settings are unchanged, the operation result is as follows:

TRAINLM, Epoch 36/1000, MSE 0.00462637/0.005, Gradient 1.88134/1e-010

When the transfer function is tansig between input layer and hidden layer, the transfer function is logsig between hidden layer and output layer respectively, the other network settings are unchanged, and the operation result is as follows:

TRAINLM, Epoch 12/1000, MSE 0.00395657/0.005, Gradient 11.9806/1e-010

When the transfer function is logsig between input layer and hidden layer, the transfer function is tansig between hidden layer and output layer, the other network settings are unchanged, the operation result is as follows:

TRAINLM, Epoch 48/1000, MSE 0.00488227/0.005, Gradient 0.618599/1e-010

From the comparision ,the tansig-logsig type combination calculation has more advantage over the other three combination calculations in network convergence speed and error precision by comparing the above-mentioned combination calculations and considering the running stability. Therefore, the transfer function was set as the function tansig between input layer and hidden layer and that as the function logsig between hidden layer and output layer.

4.3.3 Training and simulation of the network

The samples without training are predicted by the trained network using the BP neural network. The paper studies 7 groups specimens, each group specimen includes 6 test data, 42 test data as a whole. The prediction was carried out by constructing the training set using 36 test data (see Table 4-1) among the above-mentioned test data and by constructing the prediction set using the other 6 test data (see Table 4-2). The parameters of the table below involves a water-binder ratio (w/b), sand ratio (s/a), maximum aggregate size (G_{max}/mm), relative crack width (w/w_0) and relative stress (σ/f_t).

The training data of tension softening curves of dam concrete by direct tension test method   Table 4-1

Number of direct tension specimens	Sample number	G_{max}	w/b	s/a	w/w_0	σ/f_t
DT[1]	1	20	0.50	0.734	0.01101	0.95811
	2	20	0.50	0.734	0.03802	0.77000
	3	20	0.50	0.734	0.06584	0.46253
	4	20	0.50	0.734	0.08015	0.40398
	5	20	0.50	0.734	0.16060	0.30599
	6	20	0.50	0.734	0.50450	0.11788

Number of direct tension specimens	Sample number	G_{max}	w/b	s/a	w/w_0	σ/f_t
DT[2]	7	10	0.50	0.797	0.01022	0.93696
	8	10	0.50	0.797	0.02634	0.68003
	9	10	0.50	0.797	0.05004	0.42950
	10	10	0.50	0.797	0.08227	0.29462
	11	10	0.50	0.797	0.13244	0.29978
	12	10	0.50	0.797	0.22864	0.13954
DT[3]	13	40	0.50	0.598	0.01046	0.84916
	14	40	0.50	0.598	0.02578	0.72403
	15	40	0.50	0.598	0.04961	0.53198
	16	40	0.50	0.598	0.08056	0.29825
	17	40	0.50	0.598	0.13213	0.22885
	18	40	0.50	0.598	0.30182	0.15748
DT[4]	19	20	0.50	0.734	0.01026	0.97592
	20	20	0.50	0.734	0.02027	0.80934
	21	20	0.50	0.734	0.03376	0.62828
	22	20	0.50	0.734	0.05102	0.46532
	23	20	0.50	0.734	0.08596	0.30237
	24	20	0.50	0.734	0.21090	0.13942
DT[5]	25	20	0.45	0.705	0.02180	0.89294
	26	20	0.45	0.705	0.03727	0.71300
	27	20	0.45	0.705	0.05366	0.55555
	28	20	0.45	0.705	0.07448	0.43185
	29	20	0.45	0.705	0.10823	0.30814
	30	20	0.45	0.705	0.21107	0.16194
DT[6]	31	20	0.35	0.65	0.01012	0.94019
	32	20	0.35	0.65	0.02640	0.81422
	33	20	0.35	0.65	0.03689	0.63080
	34	20	0.35	0.65	0.05016	0.60383
	35	20	0.35	0.65	0.09579	0.39344
	36	20	0.35	0.65	0.29463	0.12425

The validation data of tension softening curves of dam concrete by direct tension test method   Table 4-2

Number of direct tension specimens	Sample number	G_{max}	w/b	s/a	w/w_0	σ/f_t
DT[7]	1	20	0.30	0.623	0.00917	0.93372
	2	20	0.30	0.623	0.02512	0.68087
	3	20	0.30	0.623	0.05029	0.45512
	4	20	0.30	0.623	0.07851	0.36602
	5	20	0.30	0.623	0.13318	0.24858
	6	20	0.30	0.623	0.29362	0.15712

4.3 Prediction of the Tension Softening Curve of Dam Concrete

The network structure was 4-9-1, target error was 0.005 and the learning rate was 0.1. The selected network is the better prediction network by a lot of pilot calculations. The test data of the 6 groups specimens from DT[1] to DT[6] were taken as the training data, then the softening curve of the 7th group specimen DT[7] was predicted and verified. The data training is as follows:

P=[20 0.5 0.734 0.01101;20 0.5 0.734 0.03802;20 0.5 0.734 0.06584;
20 0.5 0.734 0.08015;20 0.5 0.734 0.16060;20 0.5 0.734 0.50450;
10 0.5 0.797 0.01022;10 0.5 0.797 0.02634;10 0.5 0.797 0.05004;
10 0.5 0.797 0.08227;10 0.5 0.797 0.13244;10 0.5 0.797 0.22864;
40 0.5 0.598 0.01046;40 0.5 0.598 0.02578;40 0.5 0.598 0.04961;
40 0.5 0.598 0.08056;40 0.5 0.598 0.13213;40 0.5 0.598 0.30182;
20 0.5 0.734 0.01026;20 0.5 0.734 0.02027;20 0.5 0.734 0.03376;
20 0.5 0.734 0.05102;20 0.5 0.734 0.08596;20 0.5 0.734 0.21090;
20 0.45 0.705 0.02180;20 0.45 0.705 0.03727;20 0.45 0.705 0.05366;
20 0.45 0.705 0.07448;20 0.45 0.705 0.10823;20 0.45 0.705 0.21107;
20 0.35 0.65 0.01012;20 0.35 0.65 0.02640;20 0.35 0.65 0.03689;
20 0.35 0.65 0.05016;20 0.35 0.65 0.09579;20 0.35 0.65 0.29463]';
t=[0.95811 0.77000 0.46253 0.40398 0.30599 0.11788 0.93696 0.68003 0.42950 0.29462 0.29978
0.13954 0.84916 0.72403 0.53198 0.29825 0.22885 0.15748 0.97592 0.80934 0.62828 0.46532
0.30237 0.13942 0.89294 0.71300 0.55555 0.43185 0.30814 0.16194 0.94019 0.81422 0.63080
0.60383 0.39344 0.12425];
net=newff (minmax) (P), [9 1], {'tansig','logsig'},'trainlm');
net.trainParam.epochs=1000;
net.trainParam.goal=0.006;
LP.lr=0.1;
net=train (net,P,t);
P_test=[20 0.3 0.623 0.00917;20 0.3 0.623 0.02512;20 0.3 0.623 0.05029;
        20 0.3 0.623 0.07851;20 0.3 0.623 0.13318;20 0.3 0.623 0.29362]';
Y=sim (net,P_test)
operation results: Y = 0.9045   0.8022   0.5840   0.3918   0.2467   0.2020
original data Y= 0.93372   0.68087   0.45512   0.36602   0.24858   0.15712

The predicted data is close to the original data. The softening curve of the specimen DT [7] was achieved and shown in the Fig. 4-3.

P=[0;0.00917;0.02512;0.05029;0.07851;0.13318;0.29362;1]';
t=[1 0.9045 0.8022 0.5840 0.3918 0.2467 0.2020 0];
plot (P,t);

The softening curve by direct tension test method was compared with that obtained by neural network prediction. The comparison result is shown as Fig. 4-4:

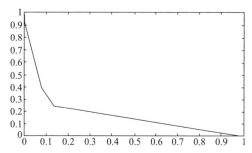

Fig. 4-3　The softening curve of the 7th group specimen by BP algorithm

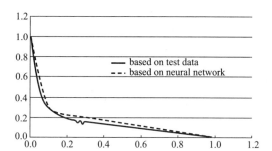

Fig. 4-4　Comparison between the softening curve by direct tension test method and that by neural network prediction

4.4　Conclusions

This paper predicted the tension softening curves of dam concrete within a certain range based on the test data of direct tension tests of dam concrete by employing the BP neural network. This approach can accelerate to get the softening relationships for the same component material but different mixtures dam concrete and save test cost of the complex and expensive direct tension tests of dam concrete. This approach have great prospects for development on scientific research and application of large - scale hydraulic engineering.

To improve the quality and extent of the training sample of BP neural network can contribute to improve the prediction. The four influencing factors water-binder ratio, sand ratio, maximum aggregate size and relative crack width (w/w_0) were considered to construct the training set. The precision of prediction can be further enhanced if the network input parameters, such as the aggregate type, the cement grade, and the type and content of the admixture are increased.

References

[1] Dariusz Alterman, Hiroshi Akita, Janusz Kasperkiewicz:Tension softening curves described by algebraic formulas and artificial neural networks, Advances in Construction Materials 2007,Part IX, p.723

[2] Hou-Gui Zhou, Qing-Bin Li and Zhi-Fang Zhao: Research on fracture simulation of mass concrete cracks. Cooperative Research Report of China Gezhouba(Group) Corporation.,Tsinghua University and Yantai University (2004) [In Chinese]

[3] Zhifang Zhao, Bo Pang, Zhigang Zhao: Fracture Behaviors of Dam and Wet-screening Concrete by Direct

Tension Test. Key Engineering Materials, Vols. 400, p.233 (2009)

[4] ShuangCong:The neural network theory and application to Matlab toolbox (Press of University of Hefei Science and Technology of China 2003),p.63[In Chinese]

[5] KuiDai:The realization technology of Neural network(Press of University of Defense Science and Technology 1998) ,p.303[In Chinese]

5 Effect of Specimen Size on Fracture Energy and Softening Curve of Concrete: Part I. Experiments and Fracture Energy

5.1 Introduction

Fracture energy is a fundamental fracture parameter, representing cracking resistance and fracture toughness of concrete, and is generally considered as a material property in concrete fracture mechanics and cracking analyses. However, it remains controversial as to whether or not the fracture energy is size dependent. Existing studies[1-5] have shown that the experimentally determined fracture energy increases with increased specimen size. Some researchers have argued that the fracture energy is a material property, and the observed size effect is caused by several sources of experimental error, the testing procedure or limitations of the three-point bend test in finding the fracture energy[6-10]. In practice, three-point bend tests are performed in most studies on size effect of fracture energy; however, this testing method could entail considerable experimental error, as noted in the literature[7-9]. Hu and Wittmann[11] here stated that the size effect is caused by variation in the width of the fracture process zone according to the ligament length of the specimen.

In order to more clearly verify the effect of specimen size and geometry on the fracture energy, three-point bend tests for a notched beam and wedge splitting tests were performed simultaneously for thirty-four specimens comprised of ten different concrete mixes, where the maximum size of coarse aggregate and the water to binder ratio were varied. The load-deflection and the load-CMOD measured from the companion specimens were averaged by a data processing method suggested in this study, and the fracture energy for each specimen was calculated from the averaged data. From the calculated fracture energy for every specimen, the effects of specimen size, geometry, aggregate size, and water to binder ratio on the fracture energy are analyzed and discussed.

5.2 Experiments

5.2.1 Materials

Table 5-1 shows the concrete mix proportions. There are three groups of concrete mixes; SG group is concrete made of small size gravel, LG group is dam concrete, and WG group is wet-screening concrete where the gravel of more than 40mm is removed through screening the fresh mixes, LG1 and LG2, respectively. Type IV cement, fly ash, crushed gravel, and river sand were used in all the mixes. The water to binder ratio was ranged from 0.30 to 0.50, and the gravel size from 10mm to 80mm. The substitution rate of fly ash was 10%-30% for the total binder content.

Mix proportions — Table 5-1

Concrete mix		w/b	Maximum gravel size	Unit weight (kg/m^3)						
				c	w	F.A	S	G	W.R.	A.E
SG	1	0.50	10mm (0.4in.)	196	140	84	869	1090	1.68	0.0196
	2	0.50	20mm (0.8in.)	185	132	79	846	1152	1.58	0.0185
	3	0.45	20mm (0.8in.)	240	135	60	814	1154	1.80	0.0195
	4	0.35	20mm (0.8in.)	309	135	77	744	1145	2.32	0.0251
	5	0.30	20mm (0.8in.)	420	140	47	698	1121	2.80	0.0280
	6	0.50	40mm (1.6in.)	168	120	72	769	1287	1.44	0.0168
LG	1	0.45	80mm (3.1in.)	159	102	68	625	1496	1.36	0.0159
	2	0.50	80mm (3.1in.)	143	102	61	653	1491	1.22	0.0143
WG	1	0.45	40mm (1.5in.)	159	102	68	625	1496	1.36	0.0159
	2	0.50	40mm (1.5in.)	143	102	61	653	1491	1.22	0.0143

SG; small gravel concrete, LG; large gravel concrete, WG; wet-screening concrete.

5.2.2 Test program

Table 5-2 shows the test program. Both three-point bend tests for a notched beam and wedge splitting tests were performed for the six mixes, SG1 to SG6. The beam tests were carried out for LG1 and WG1, and the wedge splitting tests for LG2 and WG2. The mixes SG1, SG2, LG1, LG2, WG1, and WG2 have from two to five different size specimens for the beam and wedge splitting tests. The tests were designed so as to allow for comparison of the test results according to the specimen size, maximum aggregate size, water to binder ratio, and testing method for ten different concrete mixes.

Test program — Table 5-2

Concrete mixes	Test method	
	Beam	Wedge
SG1	○	○
SG2	○	○
SG3	○	○
SG4	○	○
SG5	○	○

5 Effect of Specimen Size on Fracture Energy and Softening Curve of Concrete: Part I. Experiments and Fracture Energy

	Test method	continue
Concrete mixes	Beam	Wedge
SG6	○	○
LG1	○	X
LG2	X	○
WG1	○	X
WG2	X	○

5.2.3 Specimen and test set-up

Fig. 5-1 shows the geometry of the beam and wedge specimens, and the dimensions are listed in Tables 5-3 and 5-4. In addition to the fracture tests, tests for compressive strength and secant modulus were performed following the ASTM C 469-94[12] at an age of 1 year. Compressive strength and elastic modulus were determined based on an averaged result of three identical 150mm×150mm×150mm cubes and Φ150mm×300mm cylinders, respectively. The mold for each specimen was removed 1 day after casting, and each specimen was cured by spraying water on the surface to prevent drying during a period of 28 days. All the mixes were designed for an actual dam built in China, and the specimens were kept in the same environmental condition as the dam for 1 year after curing. The characterization specimens were also stored together with the fracture specimens.

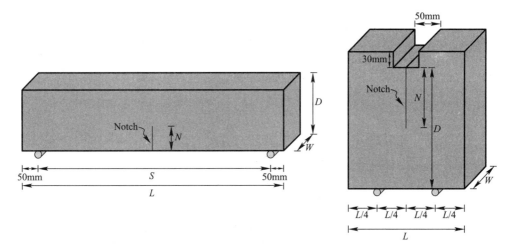

Fig. 5-1 Geometry of notched beam and wedge splitting specimens

Every test was performed at the age of 1 year. Four companion specimens were manufactured for each test variable, but some specimens were broken in handling and moving and reliable test results could not be obtained for some specimens. The numbers of companion specimens that were employed in the beam and wedge splitting tests and provided reliable data are also listed in Tables 5-3 and 5-4.

Dimensions of beam specimens Table 5-3

Concrete mix	Specimen	Number of companion specimens	Dimensions of beam specimen			
			D (mm)	L/S (mm)	N (mm)	W (mm)
SG1	B1	3	300	1300/1200	120	120
	B2	2	400	1700/1600	160	
SG2	B1	3	150	700/600	60	
	B2	2	200	900/800	80	
	B3	3	300	1300/1200	120	
	B4	2	400	1700/1600	160	
	B5	2	500	2100/2000	200	
SG3	B1	3	300	1300/1200	120	
SG4	B1	3	300	1300/1200	120	
SG5	B1	3	300	1300/1200	120	
SG6	B1	3	300	1300/1200	120	
LG1	B1	2	400	1700/1600	160	240
	B2	2	450	1900/1800	180	
	B3	4	500	2100/2000	200	
	B4	3	550	2300/2200	220	
WG1	B1	2	250	1100/1000	100	120
	B2	3	300	1300/1200	120	
	B3	3	400	1700/1600	160	

Dimensions of wedge splitting specimens Table 5-4

Concrete mix	Specimen	Number of companion specimens	Dimensions of wedge splitting specimen			
			D (mm)	L (mm)	N (mm)	W (mm)
SG1	W1	3	300	300	150	200
	W2	2	600	600	300	
SG2	W1	2	300	300	150	
	W2	3	600	600	300	
	W3	3	800	800	400	
	W4	3	1000	1000	500	
SG3	W1	3	300	300	150	
SG4	W1	2	300	300	150	
SG5	W1	3	300	300	150	
SG6	W1	3	300	300	150	

5 Effect of Specimen Size on Fracture Energy and Softening Curve of Concrete: Part I. Experiments and Fracture Energy

continue

Concrete mix	Specimen	Number of companion specimens	Dimensions of wedge splitting specimen			
			D (mm)	L (mm)	N (mm)	W (mm)
LG2	W1	3	450	450	225	250
	W2	2	800	800	400	
	W3	3	1000	1000	500	
WG2	W1	2	300	300	150	200
	W2	2	600	600	300	
	W3	3	800	800	400	

Fig. 5-2(*a*) shows the set-up for the three-point bend beam test. The tests were conducted in a very stiff servo-hydraulic closed-loop testing machine. A 100kN capacity load cell was used to measure the applied load, and the accuracy was ±2% of the maximum applied load. The crack mouth opening displacement (CMOD) was measured with a displacement sensor having a capacity and accuracy of 5mm and ±0.0005mm, respectively. The vertical deflection was also measured at the loading point. The loading actuatorwas controlled by a constant CMOD rate of 0.15mm/min.

Some wedge splitting specimens were much larger than ordinary size, because large size coarse aggregate was used. One line support is generally located in the center of the wedge specimen[13-15]. In contrast with an ordinary wedge splitting test, two line supports located in the center of the half section of the specimen were used in order to circumvent difficulties in handling the very large size specimen and to prevent unexpected failure of the specimen while preparing the test. As shown in Fig. 2(*b*), two massive steel loading devices equipped with roller bearings on each side were placed on top of the specimen. A steel profile with two identical wedges was fixed at the upper plate of the testing machine. The wedges enter between the bearings, which apply a horizontal splitting force (P_S). The displacement sensor and load cell were identical to those used in the beam test. The axis of the roller was aligned with the displacement sensor, that is, the displacement was measured at the horizontal loading axis. The wedge splitting test was also controlled by a constant CMOD rate of 0.15 mm/min. Fig. 5-3 shows the real test set-up.

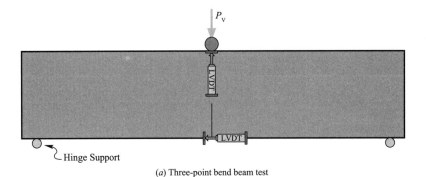

(*a*) Three-point bend beam test

Fig. 5-2 Configuration of test set-ups (1)

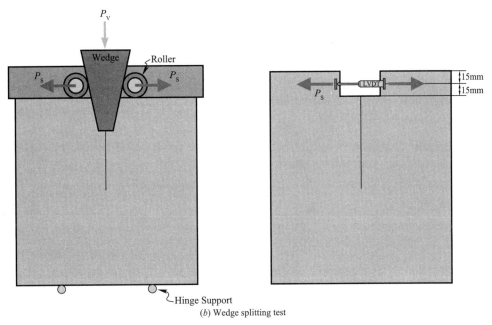

(b) Wedge splitting test

Fig. 5-2　Configuration of test set-ups (2)

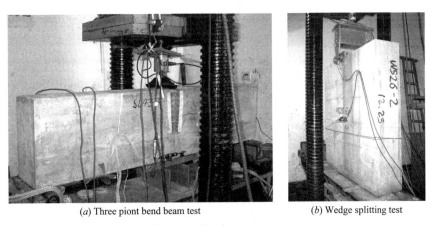

(a) Three piont bend beam test　　　(b) Wedge splitting test

Fig. 5-3　Real test set-ups

To make the notch in every specimen, steel plate with the thickness of 2mm was fixed in the molds before casting, and the surfaces facing concrete were painted with lubricating oil to prevent friction between the plate and concrete. When demolding, the steel plate was carefully removed.

5.3　Data Processing for Companion Specimens

5.3.1　Averaging the data for companion specimens

There were two to four companion specimens for each test. In order to find the fracture characteristics for each test, it is necessary to average the test results for the companions. However, the averaged results may depend on the method used to average the data. The method used to process the test results for the companion specimens is described below.

In Fig. 5-4, the data processing procedure for three companion specimens of LG2-W1 is illustrated as an example for beam andwedge splitting specimens. Fig. 5-4(a) shows the raw test data for three companion specimens, where the total number of raw data for each specimen was about 50,000. The data scattered far from the load-CMOD curve are first filtered and then one point is taken every 20 points in the load-CMOD curve. Around a given point, an average was taken of five points consisting of the given point, and two points above and below the given point. By averaging the five adjacent points, the effect of fluctuation in the measurement can be emoved. Fig. 4(b) shows the results after filtering and averaging for each companion. The peak load point of each specimen are taken from the results given in Fig. 5-4(b), and then 100 equally spaced CMOD values are calculated from the zero point to the CMOD at the peak point and from the peak to the end point, respectively. The end point of CMOD for each specimen is set such that the distance from the peak to the end point is identical for each three specimens, for example, 3.8mm from the CMOD at the peak in the case of the LG2-W1 specimen. The load values corresponding to the 100 CMOD values are calculated by interpolation between the averaged data points of Fig. 5-4(b) for ascending and descending parts. Now, we have two hundreds data points, one hundred each for ascending and descending parts. The three ascending curves for each of the three specimens(LG2-W1-1, LG2-W1-2 and LG2-W1-3) are shown in Fig. 5-4(c) and Fig. 5-4(d).

According to the equivalent elastic crack approach based on the Griffith-Irwin cracking mechanism[16-18], the fracture process zone (FPZ) is formed with an increase of the applied load, and the equivalent extension of the crack length, which is elastically equivalent to the process zone, is largest at the peak. After the peak load, the process zone starts to move forward along the crack path, maintaining the largest equivalent extension size c, as shown in Fig. 5-5. When averaging the test data between the companions, each averaging point of the data has to represent the same fracture status. The 100th data point represents the peak load status for each specimen, and is in the same fracture status. For each of the equally spaced 100 CMODs up to the peak and the corresponding loads, the ith data point for each specimen can be assumed to be in the same status, namely, the ith step of increasing the fracture process zone. The end points (200th data points) of the descending part in Fig. 5-4(d) represent the same status that the largest equivalent extension is moved the same distance from the peak point. The ith data point in the descending region can also be considered as the same status that the fracture process zone is equally far from the peak, because each data point is equally spaced. Therefore, 200 equally spaced CMOD values and the corresponding load values for each specimen can be averaged. The averaged load-CMOD curve is shown in Fig. 5-4(e).

5.3.2 Extracting the data points representing the load-CMOD curve

The data sets are prepared for an inverse analysis to find the softening curve, which will be presented in Part II. Because the number of data points used in optimal fitting significantly affects the computer running time, the minimum number of data points sufficiently representing the load-CMOD curve is desirable. From the averaged data for the companion specimens, the minimum data points were extracted using an optimization technique.

A multi-linear function was employed to fit the ascending part of Fig. 5-4(f). The points comprising the multi-linear lines were assumed to be on the line interpolated between the averaged data points. Several multi-linear functions were tried, and it was found that the multilinear function consisting of seven lines accurately fit the ascending part. With more than seven lines, the accuracy was almost identical. The Marquardt-Levenberg method[19] was used in the optimization, and the coefficient of determination was more than 0.99 for all cases. Six points excluding the origin and the peak point were found for the ascending part of every beam and wedge splitting specimen.

For the descending part, the data points equally spaced over the CMOD axis were extracted. Minimum twelve points were needed for an accurate fit. Twelve points excluding the peak and the end point points were selected, and the coefficient of determination for the multi-linear function consisting of the selected points was more than 0.99. A total of twenty points including the peak and the end pointwere extracted from the averaged load-CMOD curve, as shown in Fig. 5-4(f).

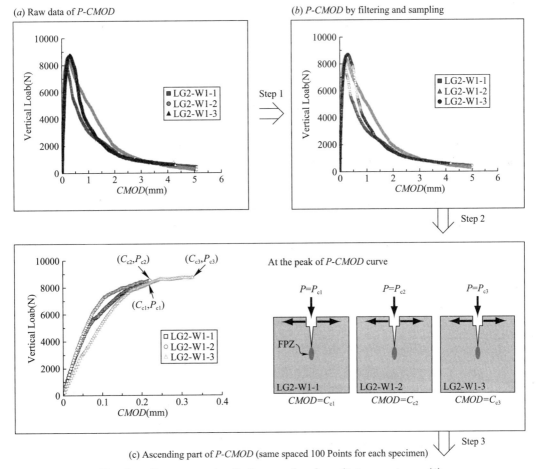

Fig. 5-4 Data processing for beam and wedge splitting specimens (1)

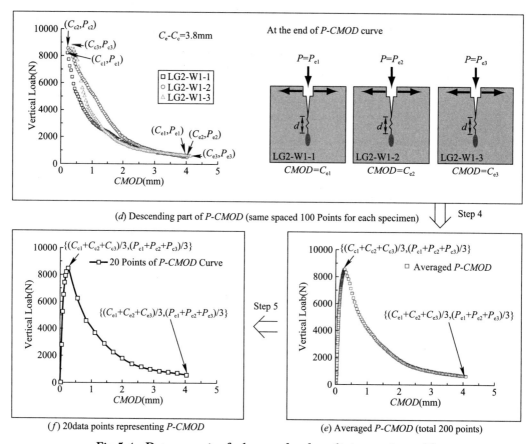

Fig. 5-4 Data processing for beam and wedge splitting specimens (2)

5.4 Test Results and Discussion

Table 5-5 shows the compressive strength, the secant modulus, and the tangent modulus at 1 year age for every concrete mix. Because the inelastic strain is included when calculating the secant modulus, the initial tangent modulus excluding the inelastic part was calculated by the following equation of the CEB-FIP model code[20], which will also be used in the inverse analysis of Part II.

Mechanical properties　　　　Table 5-5

Concrete mix	Compressive strength (MPa)	Secant modulus (GPa)	Initial tangent modulus (GPa)
SG1	43.8	26.7	31.4
SG2	43.4	33.3	39.2
SG3	50.9	30.4	35.7
SG4	56.4	30.5	35.9
SG5	50.2	34.8	41.0

5.4 Test Results and Discussion

continue

Concrete mix	Compressive strength (MPa)	Secant modulus (GPa)	Initial tangent modulus (GPa)
SG6	50.8	33.1	38.9
LG1	40.0	28.6	33.6
LG2	51.7	33.7	39.6
WG1	40.0	28.6	33.6
WG2	51.7	33.7	39.6

$$E_{ci} = E_c/0.85 \tag{5-1}$$

In the equation, E_c and E_{ci} are the secant and tangent modulus, respectively. The averaged load-deflection curves for the beam specimens are shown in Fig. 5-6, 5-7 and 5-8 showthe load-CMOD curve consisting of 20 data points (7 points for ascending and 13 points for descending) corresponding to Fig. 5-4(f) for every beam and wedge specimen.

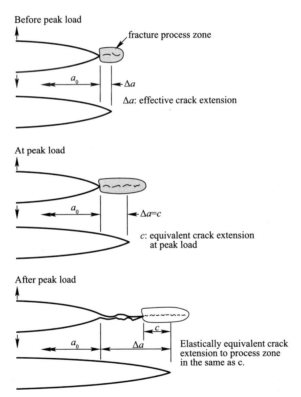

Fig. 5-5 Fracture process zone and equivalent crack extension at different fracture stages[18]

Peak loads measured from the beam and wedge splitting specimens are compared in Fig. 9. The peak loads of the beam and wedge specimens, SG1, SG2, LG1, WG1, LG2 and WG2, increase with increasing the specimen size, as shown in Fig. 5-9(a) and Fig. (b). For the same size beam specimens, SG1-B1, and SG3-B1 to SG6-B1, the peak loads of SG1 and SG5 are the lowest and the largest, respectively. Similarly, for the same size wedge specimens, SG1-W1,

5 Effect of Specimen Size on Fracture Energy and Softening Curve of Concrete: Part I. Experiments and Fracture Energy

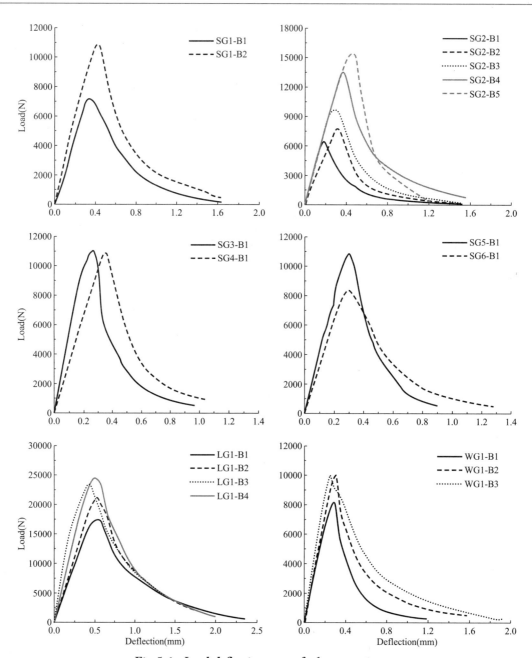

Fig. 5-6　Load-deflection curves for beam specimens

SG2-W1, and SG3-W1 to SG6-W1, the peak loads of SG1 and SG5 are also the lowest and the largest, respectively. This appears to be attributable to the effect of the water:binder ratio.

The fracture energy for each specimen was calculated from the test results and is compared according to size in Figs. 5-10~5-12. In the case of the beam specimen, the fracture energy was calculated based on the following equation of Rilem recommendation[21], and the self weight effect was also considered.

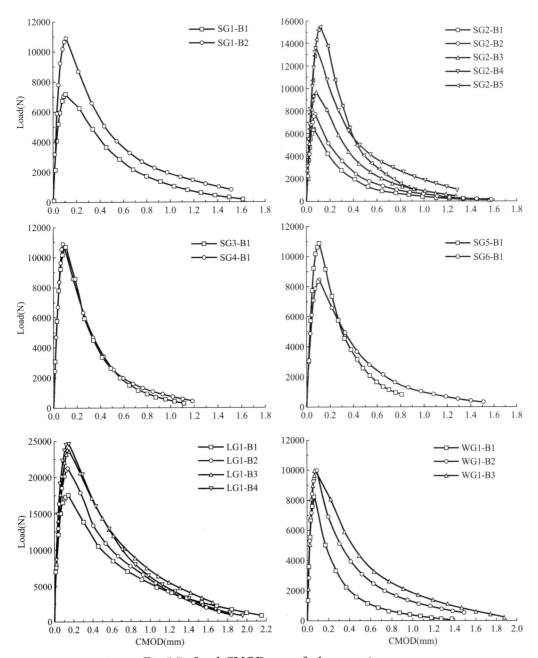

Fig. 5-7 Load-CMOD curves for beam specimens

$$G_F = \frac{W_1 + mg\delta_0}{(D-N)W} \qquad (5\text{-}2)$$

In Eq. (5-2), G_F is the fracture energy, W_1 is the area under the loaddeflection curve of Fig. 6, m is the total mass of the specimen, g is gravity, δ_0 is the end deflection at $P=0$, and D, N, and W are the depth, notch length, and width of the specimen, respectively. The fracture energy for the wedge *specimen* was calculated from the load-CMOD of Fig. 5-7 using the following equation; note that the self weight of

5 Effect of Specimen Size on Fracture Energy and Softening Curve of Concrete: Part I. Experiments and Fracture Energy

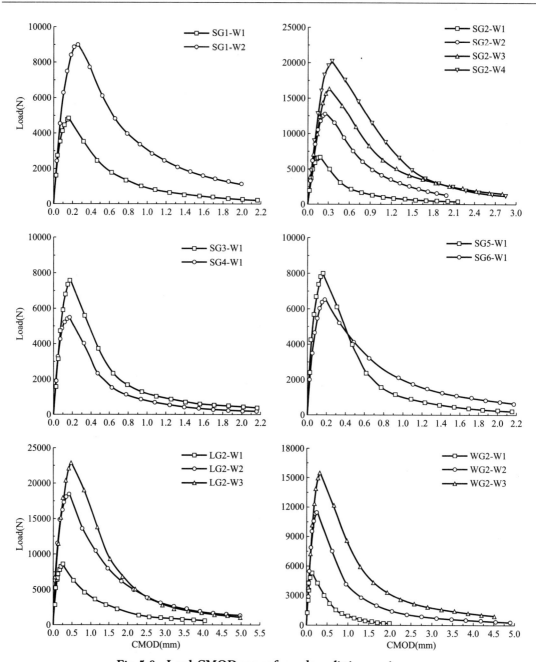

Fig. 5-8 Load-CMOD curves for wedge splitting specimens

the wedge specimens was not considered, because the selfweight is borne by the two bottom supports, as shown in Fig. 5-2(b), and is balanced to the reaction at the supports.

$$G_F = \frac{W_1}{(D-N)W}, \quad P_s = \frac{P_v}{2\tan\theta} \tag{5-3}$$

In Eq. (5-3), W_1 is the total work of the area under the load-CMOD curve of Fig. 5-8, P_s is the horizontal load, P_v is the applied vertical load, and θ is the wedge angle, which was 15° in the

5.4 Test Results and Discussion

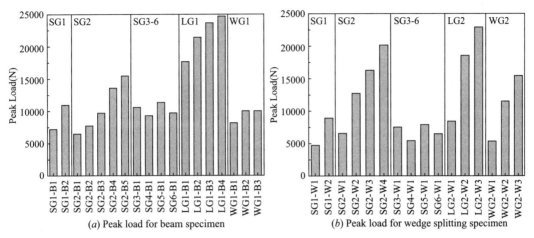

(a) Peak load for beam specimen

(b) Peak load for wedge splitting specimen

Fig. 5-9 Maximum load

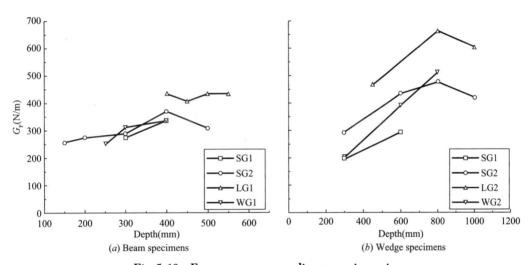

(a) Beam specimens

(b) Wedge specimens

Fig. 5-10 Fracture energy according to specimen size

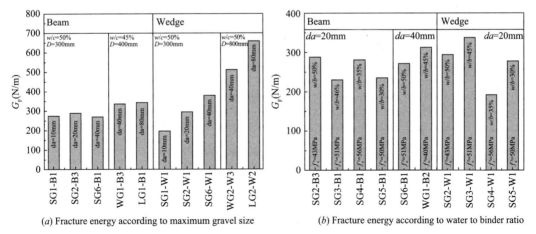

(a) Fracture energy according to maximum gravel size

(b) Fracture energy according to water to binder ratio

Fig. 5-11 Fracture energy according to gravel size and w/b ratio

59

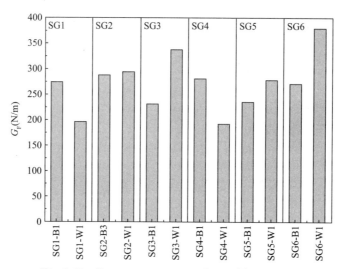

Fig. 5-12 Fracture energy according to Test method

test set-up. The tests did not proceed until a full tail of the load-deflection curve and the load-CMOD curve, respectively, was obtained, and the tail part that is not measured was assumed to be a straight line with the last slope of the load-deflection and load-CMOD curves in the calculation of fracture energy.

The fracture energy calculated from the different size beam specimens is plotted over their depth in Fig. 5-10(a). It is seen that the fracture energy increases slightly with size in SG1, SG2, and WG1 concrete, and there is no increasing tendency with size in LG1. The difference between the maximum and the minimum fracture energies was 20%-30% of the maximum value.

The sources of error in the calculation of the fracture energy from the beam test have previously been investigated by Guinea et al.[7], Planas et al.[8], and Elices et al.[9]. The main sources were the friction between the support and the specimen, crushing near the supports, the energy dissipation in the bulk of the materials near the crack, and the energy enclosed in the unmeasured tail part of the load-deflection curve. It was found in these studies that the observed size dependence of the fracture energy of about 40% was caused by these sources, and the size-independent fracture energy could be obtained from correction of these errors. The majority of the error, more than 30% of the observed size dependence of 40%, originated from the neglected energy of the tail part. When these sources of error are considered in the calculation of the fracture energy for the beam tests of this study, it is not clear whether the 20%-30% size dependence of Fig. 5-10(a) is an intrinsic feature of concrete material or is caused by inevitable test error.

Unlike in the beam tests, some experimental error can be avoided in the wedge splitting tests. Because the vertical load is not applied to the specimen and only lateral force acts by the wedge and roller, the energy dissipation not related to crack propagation such as friction of the support and crushing near the support can be minimized. The self weight was also compensated in the test set-up by using two bottom supports. Therefore, if the size dependence observed in the beam tests is due to an artefact, there would be no or less size dependence in the fracture energy obtained from the wedge splitting tests for the same concrete. As shown in Fig. 5-10(b), however, the fracture energy obtained from the wedge specimens also increases over size, and the difference between the maximum and the minimum fracture energies was 30%-60% of the maximum value (140%-250% of the minimum value). It is seen that the size dependence in the wedge splitting tests is much larger than that in the beam tests. Even if the sources noted above are considered, it is difficult to explain this amount of size dependence. Therefore, it can be concluded that the fracture energy of concrete depends on the specimen size.

In a previous study by Wittmann et al.[2], wedge splitting tests were performed with three different size specimens, and the size effect of the fracture energy, i.e., a maximum of 30% fracture energy difference for one to four times size specimens, has been observed. They also attempted to find the softening curve by optimally fitting the test results using an optimization technique, and the influence of the size dependent fracture energy on the softening curve was analyzed. However, this study was conducted only for one concrete and thus more extensive investigation of the size effect of the softening curve is needed (this will be carried out by the present authors in our subsequent paper).

The size effect of fracture energy can be explained in terms of variation of the influence of the fracture process zone on fracturing according to the size and geometry of the specimen. This is because the whole body except the process zone can be assumed to be elastic in the beam and wedge splitting tests, although some regions near the support and the loading point are in an inelastic state, but do not contribute to cracking. Hu and Wittmann[11] introduced a local fracture energy concept where the fracture energy varies with the width of the process zone. They explained the size dependent fracture energy in terms of different distributions of the local fracture energy according to the shape of the process zone, which depends on the size and geometry of the specimen. However, the mechanism of the process zone is still not fully understood.

The fracture energy of SG2 and LG2 increases up to the second largest size and drops slightly at the largest size, as shown in Fig. 5-10(a) and (b). It appears that the fracture energy does not decrease but rather it approaches a certain value with an increase of the size. This asymptotic behaviour was also observed in Wittmann, Mihashi and Nomura's experimental work[12]. Further study on the mechanism of the process zone inducing the increase of fracture energy and the asymptotic feature is needed.

The beam and wedge splitting tests were performed for ten different concretes, and the fracture energies were analyzed according to the maximum aggregate size and the water to binder ratio, as presented in Fig. 5-11, where d_a is the maximum aggregate size. Because of the possible size effect, the fracture energies for the same size specimens were compared. In Fig. 5-11(a), the effect of the gravel size does not clearly appear in the beam tests, but the wedge test results distinctly show the effect. As noted above, the beam test results may include more experimental error compared to the wedge splitting test, which seems to hinder the effect on the gravel size in the beam tests. The fracture energies of WG2-W3 and LG2-W2 are much higher than others. In these cases, the specimen sizes (800mm depth) are more than double those of the other specimens. The effect of gravel size on fracture energy found in this study is in accordance with the findings of the literature[22]. As shown in Fig. 5-11(b), there is no increasing or decreasing trend in the fracture energy for the water to binder ratio. The values given at the bottom of the bar are the measured compressive strength and the fracture energy is for the same depth of 300m, as presented in Fig. 5-11(b). Generally, the fracture energy increases with an increase of the compressive strength, which strongly depends on the water to binder ratio. However, the substitution ratio of fly ash to the cement was different in each mix. There was little difference in the strength in SG3 to SG5, although the water to binder ratio varied from 30% to 45%. Therefore, the water to binder ratio apparently cannot be directly correlated to the fracture energy in this study.

The fracture energies obtained from the beam and wedge splitting tests for SG1 to SG6 concrete are compared in Fig. 5-12. Because the possible energy dissipation in regions other than the fracture process zone, the experimentally determined fracture energy from the beam test is expected to be larger than that of the wedge test. However, there is no systematic trend between the beam and wedge geometry. Because of the different stress and strain distribution according to the geometry, the shape of the process zone might be affected by the geometry of the specimen as well as the aggregate size and the mix proportion of the matrix[17]. In order to explain the difference of the fracture energies obtained from the two tests, every influencing factor on the process zone, including the specimen size and geometry, the aggregate size, and the mechanical properties of the matrix, should be considered.

A qualitative analysis was performed for the experimentally determined fracture energy. In the subsequent paper, the size effect of the fracture energy will be quantitatively examined based on the softening curves obtained from an inverse analysis, and a possible mechanism of the process zone related to the identified features for the fracture energy will be discussed.

5.5 Conclusions

Three-point bend tests for a notched beam and wedge splitting tests were performed for ten different types of concrete with different size specimens. A data averaging method for the companion specimens in the fracture test was proposed, and the fracture energy was calculated from the averaged data for each specimen. From the comparison of the fracture energies, it was found that the fracture energy increases with an increase of the specimen size, especially in the wedge tests, and asymptotic behaviour over the size is observed in some concretes. Additionally, it was shown that the fracture energy increases with an increase of the maximum aggregate size, but there was no systematic trend with the water to binder ratio and the test method. The observed size effect of the fracture energy seems to be attributed to the different shape of fracture process zone according to the specimen size and geometry. Further study is needed to identify the mechanism of the process zone. The effect of the size dependent fracture energy on the softening curvewill be analyzed froman inverse analysis, and a possible mechanism related to the size effect will be discussed in Part Ⅱ.

References

[1] A. Hillerborg, Results of three comparative test series for determining the fracture energy G_F of concrete, Materials and Structures 18 (107) (1985) 407-413.

[2] F.H. Wittmann, H. Mihashi, N. Nomura, Size effect of fracture energy of concrete, Engineering Fracture Mechanics 35 (1990) 107-115.

[3] S.P. Shah, S.E. Swartz and C.S. Quyang. Fracture mechanics of concrete: applications of fracture mechanics to concrete rock and other quasi-brittle materials. Wiley, New York, 1995.

[4] P. Maturana, J. Planas, M. Elices, Evolution of fracture behavior of saturated concrete in low temperature range, Engineering Fracture Mechanics 35 (1990) 827-834.

[5] S. Xu, G. Zhao, Fracture energy of concrete and its variational trend in size effect studied by using three point bending beams, Journal of Dalian University of Technology 31 (1) (1991) 79-86.

[6] X.H. Guo, R.I. Gilbert, The effect of specimen size on the fracture energy and softening function of concrete, Materials and Structures 33 (2000) 309-316.

[7] G.V. Guinea, J. Planas, M. Elices, Measurement of the fracture energy using three point bend tests: part 1—influence of experimental procedures, Materials and Structures 25 (1992) 212-218.

[8] J. Planas, M. Elices, G.V. Guinea, Measurement of the fracture energy using three point bend tests: part 2-influence of bulk energy dissipation, Materials and Structures 25 (1992) 305-312.

[9] J. Planas, M. Elices, G.V. Guinea, Measurement of the fracture energy using three point bend tests: part 3—influence of cutting the P-d tail, Materials and Structures 25 (1992) 327-334.

[10] Q. Jueshi, L. Hui, Size effect on fracture energy of concrete determined by three point bending, Cement and Concrete Research 27 (7) (1997) 1031-1036.

[11] X.Z. Hu, F.H. Wittmann, Fracture energy and fracture process zone, Materials and Structures 25 (1992) 319-326.

[12] ASTM C 469-94, standard test method for static modulus of elasticity and Poisson's ratio of concrete in compression, Annual Book of ASTM Standards, West Conshocken, Pa, 1994.

[13] J.K. Kim, Y. Lee, S.T. Yi, Fracture characteristics of concrete at early ages, Cement and Concrete Research 34 (2004) 507-519.

[14] H.N. Linsbauer, E.K. Tschegg, Fracture energy determination of concrete with cubeshaped specimens, Zement and Beton 31 (1986) 38-40 (in German).

[15] E. Brühwiler, F.H. Wittmann, The wedge splitting test, a new method of performing stable fracture mechanics tests, Engineering Fracture Mechanics 35 (1/2/3) (1990) 117-125.

[16] Y.S. Jenq, S.P. Shah, A two parameter fracture model for concrete, Journal of Engineering Mechanics 111

(4) (1985) 1227-1241.

[17] Z.P. Bazant, M.T. Kazemi, Determination of fracture energy, process zone length and brittleness number from size effect with application to rock and concrete, International Journal of Fracture 44 (1990) 111-131.

[18] Z.P. Bazant, M.T. Kazemi, Size effect in fracture of ceramics and its use to determine fracture energy and effective process zone length, Journal of American Ceramic Society 73 (7) (1990) 1841-1853.

[19] K.M. Brown, Derivative-free analogues of the Levenberg. Marquardt and Gauss algorithms for non-linear least square approximations, IBM, Philadelphia Scientific Center Technical Report, No. 320-2994, 1970.

[20] Comite Euro-International Du Beton, CEB-FIP Model Code 1990, Thomas Telford, 1993.

[21] Rilem TC-50 FMC (Draft Recommendation), Determination of the fracture energy of mortar and concrete by means of three point bend tests on notched beams, Materials and Structures 18 (106) (1985) 285-290.

[22] Szczepan Wolinski, Dirk A. Hordijk, Hans W. Reinhardt, Hans A.W. Cornelissen, Influence of aggregate size on fracture mechanics parameters of concrete, International Journal of Cement Composites and Lightweight Concrete 9 (2) (1987) 95-103.

6 Effect of Specimen Size on Fracture Energy and Softening Curve of Concrete: Part II. Inverse Analysis and Softening Curve

6.1 Introduction

A softening curve is representing the relationship between the crack opening displacement and the gradual stress drop after the tensile strength. It is commonly used in the simulation of tensile cracking of concrete. The crack band model[1] and cohesive crack model[2] are representative approaches in practical finite element analyses for cracking. In the former, the softening behavior is incorporated with the elastic behavior in an element by scaling the softening curve according to the element size. In the latter, the crack path coincides with boundaries between elements, and the softening curve directly represents the relationship between transverse displacement of the boundaries and the corresponding cohesive force. One advantage of the softening curve is that it fully describes the fracture process zone (FPZ). However, the fracture process zone is simplified: it is lumped into a line in the cohesive crack model, and its inelastic deformation is smeared over a band of elements exhibiting softening behavior in the crack band model. The lattice model[3,4] and nonlocal damage model[5] are among some of the approaches that enable a more realistic simulation of the FPZ, including its width and length. Although these methods are useful for understanding the fracture process and determining the interactions between the characteristics in different scales, there are limitations in computational effectiveness, three-dimensional modeling, and predictive capability.

In this study, softening curves for every specimen assessed in the companion paper were found by inverse analyses, in which the cohesive crack model was employed and the softening curve was assumed to be a tetra-linear curve. From the analysis results, we identified how the size effect of fracture energy observed in the preceding paper[6] affects the softening curve and also examined the differences in the softening curves of the beam and wedge splitting specimens to assess if there is indeed a geometry effect. Finally, a possible mechanism for the FPZ with respect to the size effect has been discussed.

6 Effect of Specimen Size on Fracture Energy and Softening Curve of Concrete: Part II. Inverse Analysis and Softening Curve

6.2 Inverse Analysis and Softening Curve

6.2.1 General

In order to find the softening curve that optimally fits the measured load-CMOD curve, a finite element analysis and an optimization technique were incorporated. The algorithm of the inverse analysis is shown in Fig. 6-1. Fig. 6-2 shows the finite element meshes for the beam and wedge splitting specimens; the meshes were scaled up or down according to the specimen size. Only half of the specimen was modeled, considering its symmetry. The total number of elements and the number of nodes on the ligaments were 836 and 32 for the beam specimen and 601 and 36 for the wedge splitting specimen, respectively. The cohesive crack model and the principle of superposition suggested by Gopalaratnam and Ye[6] were employed to simulate the crack propagation. The stress condition was assumed as plane stress, and the initial tangent modulus, listed in the preceding paper[7], was used as a material input for the elastic body. The overall optimization procedure was the same as that employed in a previous study[8] with the exception that the Marquardt-Levenberg method[9] was used instead of the Newton-Raphson iteration method.

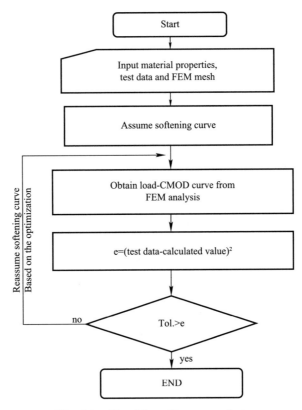

Fig. 6-1　Algorithm of inverse analysis

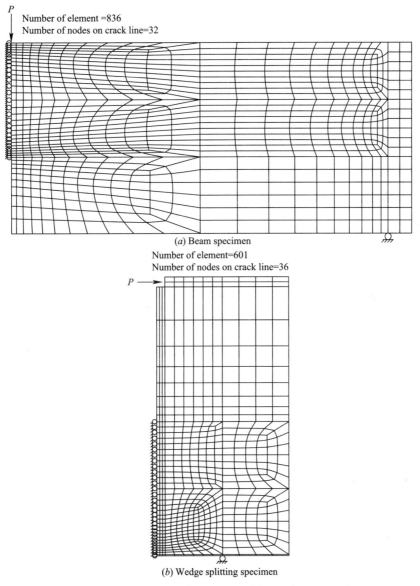

Fig. 6-2 Specimen configuration and finite element mesh

6.2.2 Procedure of inverse analysis

Fig. 6-3 shows the procedure of the inverse analysis to find the tetralinear softening curve optimally fitting the measured load-CMOD curve for each specimen. Two test data sets were prepared prior to the inverse analysis; the data extraction method was described in the preceding paper[6]. The first set consists of 20 points (8 points in the ascending part, 12 points in the descending part), and the last CMOD of the first set is almost half of the largest CMOD measured in the test. In the second set, which also consists of 20 points, the ascending part is the same as in the first set, and the descending part is roughly twice longer than that of the first set.

6　Effect of Specimen Size on Fracture Energy and Softening Curve of Concrete: Part Ⅱ. Inverse Analysis and Softening Curve

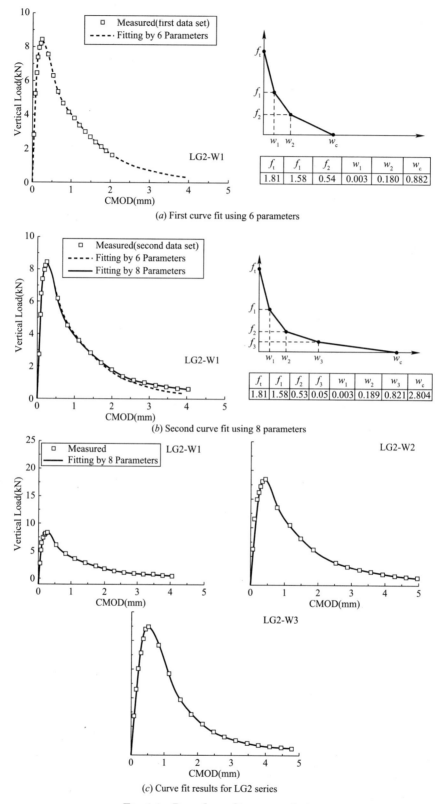

(a) First curve fit using 6 parameters

(b) Second curve fit using 8 parameters

(c) Curve fit results for LG2 series

Fig. 6-3　Procedure of inverse analysis

In the first step of the inverse analysis, the tri-linear softening curve optimally fitting the first data set was found, as shown in Fig. 7-3 (a), and six parameters of the tri-linear curve were obtained. In the second step, three parameters obtained from the first curve fit, f_t, f_1, and w_1, which do not affect the tail part of the load-CMOD curve, were fixed and the remaining five parameters among the eight parameters of the tetra-linear curve fitting the second data set were found, as shown in Fig. 6-3 (b). As shown in the left graph of Fig. 6-3(b), the tetralinear softening curve can more accurately fit the tail part of the load-CMOD curve. Fig. 6-3(c) shows the test data and the curve fit results for the LG2 series. Because of the different sensitivities of each parameter, it is difficult to directly find tetra-linear curve. In a number of studies[7,8,10], a bi-linear softening curve has been employed to simulate tensile cracking of concrete, owing to its simplicity. However, comparing the results from a previous study using a bi-linear softening curve in the optimization[8], it is found that the tetralinear curve can much more accurately fit the measured load-CMOD curve, as shown in Fig. 6-3(c).

6.2.3 Softening curves from the inverse analysis and averaging softening curves

Table 6-1 shows the eight parameters of the tetra-linear softening curve for every specimen. The eight parameters of the softening curves for different size specimens were averaged and are shown in Table 6-1. Fig. 6-4 schematically illustrates how the averaged softening curve was obtained. For example, when two softening curves are averaged, the stress is averaged at 10 points of two tetra-linear softening curves, as shown in Fig. 6-4(a). Because the first point is the tensile strength f_t, a total of nine averaged points can be obtained. The averaged curve now consists of 9 points. This was then converted to tetra-linear curve by optimization process, as shown in Fig. 6-4(b).

Parameters of tetra-linear softening curve for each specimen Table 6-1

Speci-mens			8 parameters of softening curve							
			f_t (MPa)	f_1 (MPa)	f_2 (MPa)	f_3 (MPa)	w_1 (mm)	w_2 (mm)	w_3 (mm)	w_c (mm)
Beam	SG1	B1	1.95	1.01	0.43	0.17	0.074	0.183	0.359	0.728
		B2	3.30	1.46	0.59	0.22	0.032	0.113	0.350	1.224
		Avr.	2.63	1.44	0.64	0.19	0.034	0.116	0.332	1.224
	SG2	B1	3.74	1.32	0.51	0.10	0.041	0.131	0.280	0.571
		B2	3.36	1.38	0.83	0.28	0.046	0.072	0.196	0.566
		B3	3.91	1.98	0.60	0.23	0.017	0.105	0.230	0.724
		B4	4.14	1.30	0.65	0.35	0.036	0.108	0.224	0.828
		B5	3.68	3.15	0.66	0.21	0.013	0.063	0.222	0.604
		Avr.	3.78	1.68	0.77	0.23	0.033	0.076	0.220	0.828
	SG3	B1	4.45	2.63	0.57	0.18	0.012	0.085	0.226	0.490
	SG4	B1	4.85	1.48	0.67	0.19	0.027	0.105	0.206	0.726

continue

Speci-mens			8 parameters of softening curve							
			f_t(MPa)	f_1(MPa)	f_2(MPa)	f_3(MPa)	w_1(mm)	w_2(mm)	w_3(mm)	w_c(mm)
Beam	SG5	B1	3.59	1.20	0.54	0.20	0.052	0.099	0.210	0.384
	SG6	B1	1.96	1.86	0.89	0.30	0.034	0.074	0.204	0.656
	LG1	B1	2.76	1.63	0.59	0.25	0.003	0.151	0.502	1.151
		B2	2.62	1.80	0.59	0.25	0.009	0.144	0.451	0.854
		B3	3.76	1.75	0.58	0.19	0.010	0.171	0.603	1.084
		B4	3.48	1.42	0.67	0.34	0.023	0.154	0.316	0.990
		Avr.	3.16	2.05	1.51	0.42	0.007	0.024	0.188	1.151
	WG1	B1	4.32	2.22	0.63	0.14	0.010	0.055	0.261	0.642
		B2	3.95	1.61	0.53	0.24	0.022	0.121	0.344	0.810
		B3	3.21	1.52	0.46	0.26	0.019	0.149	0.322	1.068
		Avr.	3.83	1.80	0.89	0.26	0.017	0.062	0.196	1.068
Wedge	SG1	W1	3.39	1.58	0.42	0.10	0.004	0.132	0.350	0.702
		W2	3.61	1.72	0.45	0.12	0.005	0.147	0.503	1.102
		Avr.	3.50	1.68	0.44	0.09	0.004	0.139	0.438	1.102
	SG2	W1	3.36	1.85	0.43	0.18	0.024	0.135	0.281	0.860
		W2	4.11	1.69	0.45	0.05	0.032	0.238	0.580	1.371
		W3	3.30	1.40	0.26	0.12	0.062	0.338	0.705	1.471
		W4	3.29	1.36	0.71	0.33	0.069	0.243	0.378	1.103
		Avr.	3.52	2.03	1.08	0.11	0.031	0.086	0.387	1.471
	SG3	W1	3.93	1.52	0.14	0.04	0.044	0.208	0.702	1.006
	SG4	W1	3.24	1.84	0.21	0.08	0.008	0.148	0.352	0.673
	SG5	W1	4.78	2.19	0.63	0.10	0.021	0.137	0.241	0.596
	SG6	W1	3.75	2.40	0.97	0.20	0.006	0.102	0.372	0.904
	LG2	W1	1.81	1.58	0.53	0.05	0.003	0.189	0.821	2.804
		W2	3.77	1.78	0.71	0.12	0.022	0.241	0.818	2.050
		W3	2.27	1.52	0.38	0.01	0.095	0.410	0.991	2.453
		Avr.	2.62	1.72	0.69	0.08	0.023	0.235	0.766	2.804
	WG2	W1	3.47	1.51	0.37	0.10	0.015	0.152	0.328	0.502
		W2	4.47	1.50	0.31	0.04	0.025	0.267	0.703	1.662
		W3	4.89	1.50	0.35	0.13	0.029	0.364	0.708	1.472
		Avr.	4.28	1.57	0.69	0.12	0.024	0.159	0.400	1.662

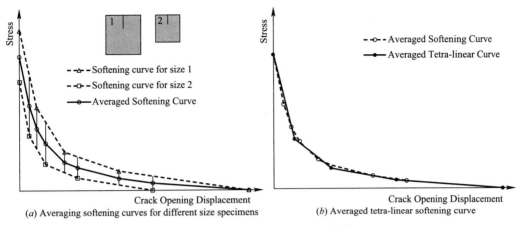

Fig. 6-4　Averaging method for softening curves of different size specimens

6.3　Analysis Results and Discussion

6.3.1　Comparison between the measured and calculated peak load and CMOD at peak

The peak loads obtained from the optimized softening curves of Table 6-1 for all the specimens are compared to the measured peak loads in Fig. 6-5, and excellent agreement between the measured and calculated results in both the beam and wedge splitting tests can be seen. The calculated and measured CMOD at peak are compared in Fig. 6, and every point is also very closely distributed to the 45° linear line. A comparison between the fracture energy calculated based on the optimized softening curve, which means the area under the softening curve, and the experimentally measured fracture energy is presented in Fig. 6-7. While the calculated and measured fracture energy are very close to each other in the wedge splitting test in Fig. 6-7 (b), the calculated values are slightly lower than the measured results in the beam test. In the case of the beam test, the measured fracture energy was obtained from the load-deflection curve and the optimized softening curve was found by fitting the load-CMOD curve. The differences among the fracture energies of Fig. 6-7(a) indicate that a part of the externally applied energy to the beam specimen was dissipated due to spurious energy loss, rather than due to the crack formation. The possible sources were noted in the preceding paper[6]. From Figs. 6-3 (c) and 6-5～6-7, it can be confirmed that the softening curves found from the inverse analysis accurately simulate the fracture response.

Peak load and corresponding CMOD for specimens of different size were calculated from the averaged softening curve and are compared to the measured values in Figs. 6-8 and 6-9. Although the scatter is larger than that of Figs. 6-5 and 6-6, good agreement between the measured and calculated values is still obtained. From the initial loading to the peak load state, the fracture process zone is not fully developed and the cohesive stress at the initial crack tip or

6 Effect of Specimen Size on Fracture Energy and Softening Curve of Concrete: Part II. Inverse Analysis and Softening Curve

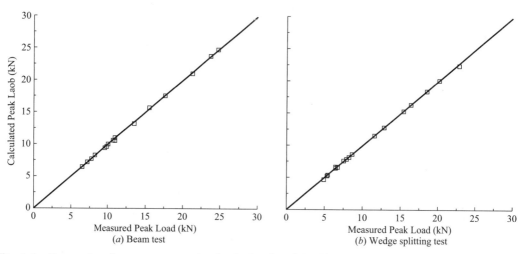

Fig. 6-5 Comparison between measured and calculated peak load (based on the optimized softening curve)

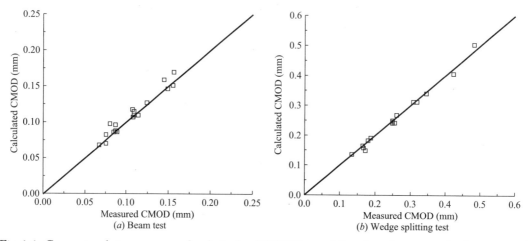

Fig. 6-6 Comparison between measured and calculated CMOD at peak (based on the optimized softening curve)

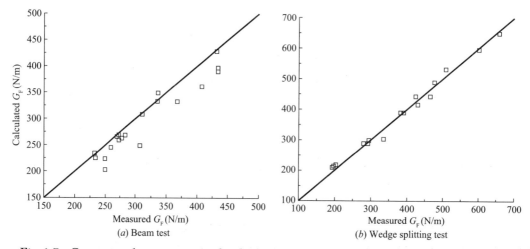

Fig. 6-7 Comparison between measured and calculated G_F (based on the optimized softening curve)

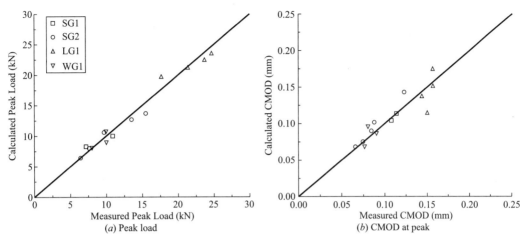

Fig. 6-8 Comparison between measured and calculated peak load and CMOD at peak for beam specimens (based on the averaged softening curve)

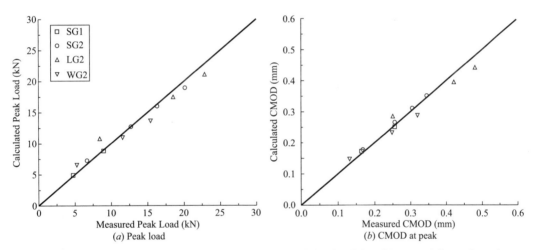

Fig. 6-9 Comparison between measured and calculated peak load and CMOD at peak for wedge splitting specimens (based on the averaged softening curve)

the end point of the initial notch remains in the first steep branch of the softening curve. Even for a specimen of infinite size, the cohesive stress at the initial crack tip is almost at the end of the first branch[11]. Therefore, the accurate fit of the averaged softening curve to the peak load and corresponding CMOD indicates that the initial steep branch of the softening curve does not depend on the specimen size; or, even if it depends on the size, the effect of the size on the initial branch is small enough to ignore. If the initial branch is similar for the different size specimens, then the remaining part of the softening curve should depend on the size to equilibrate the experimentally observed size-dependent fracture energy.

Based on the averaged curve for specimens of different size, the softening curve of each specimen was re-determined. The four stress parameters of the averaged softening curve, $f_v, f_1, f_2,$

and f_3, were fixed and the four COD parameters, w_1, w_2, w_3, w_c, optimally fitting the first found softening curve for each specimen were then found, as listed in Table 6-2.

The softening curves of the beam and wedge splitting specimens for SG3 to SG6 were also averaged using the method described in Fig. 4, and the results are listed in Table 6-3. The four COD parameters for the beam and wedge specimens, w_1, w_2, w_3, w_c, were re-determined, fixing the parameters $f_t, f_1, f_2,$ and f_3 as given in Table 6-2. The results are shown in Table 6-3.

Softening curve when stresses are fixed as averaged values   Table 6-2

	Specimens		8 parameters of softening curve							
			f_t (MPa)	f_1 (MPa)	f_2 (MPa)	f_3 (MPa)	w_1 (mm)	w_2 (mm)	w_3 (mm)	w_c (mm)
Beam	SG1	B1	2.63	1.44	0.64	0.19	0.022	0.135	0.307	0.734
		B2	2.63	1.44	0.64	0.19	0.042	0.104	0.359	1.287
	SG2	B1	3.78	1.68	0.77	0.23	0.034	0.086	0.214	0.431
		B2	3.78	1.68	0.77	0.23	0.034	0.074	0.217	0.591
		B3	3.78	1.68	0.77	0.23	0.021	0.099	0.206	0.737
		B4	3.78	1.68	0.77	0.23	0.033	0.077	0.267	0.960
		B5	3.78	1.68	0.77	0.23	0.043	0.061	0.196	0.599
	LG1	B1	3.16	2.05	1.51	0.42	0.001	0.010	0.195	1.120
		B2	3.16	2.05	1.51	0.42	0.004	0.033	0.177	0.877
		B3	3.16	2.05	1.51	0.42	0.010	0.031	0.207	1.049
		B4	3.16	2.05	1.51	0.42	0.018	0.205	0.201	0.980
	WG1	B1	3.83	1.80	0.89	0.26	0.015	0.050	0.162	0.551
		B2	3.83	1.80	0.89	0.26	0.021	0.078	0.221	0.878
		B3	3.83	1.80	0.89	0.26	0.013	0.082	0.222	1.141
Wedge	SG1	W1	3.50	1.68	0.44	0.09	0.004	0.124	0.355	0.719
		W2	3.50	1.68	0.44	0.09	0.005	0.150	0.546	1.171
	SG2	W1	3.52	2.03	1.08	0.11	0.020	0.075	0.203	1.444
		W2	3.52	2.03	1.08	0.11	0.031	0.104	0.384	1.164
		W3	3.52	2.03	1.08	0.11	0.037	0.094	0.425	1.944
		W4	3.52	2.03	1.08	0.11	0.040	0.098	0.482	2.015
	LG2	W1	2.62	1.72	0.69	0.08	0.0001	0.147	0.701	2.318
		W2	2.62	1.72	0.69	0.08	0.047	0.239	0.872	2.347
		W3	2.62	1.72	0.69	0.08	0.051	0.311	0.670	2.413
	WG2	W1	4.28	1.57	0.69	0.12	0.011	0.108	0.243	0.546
		W2	4.28	1.57	0.69	0.12	0.025	0.177	0.380	1.257
		W3	4.28	1.57	0.69	0.12	0.032	0.245	0.553	1.664

Averaged softening curves and softening curves when stresses are fixed as averaged values Table 6-3

Concrete		8 parameters of softening curve							
		f_t (MPa)	f_1 (MPa)	f_2 (MPa)	f_3 (MPa)	w_1 (mm)	w_2 (mm)	w_3 (mm)	w_c (mm)
Average	SG3	4.19	2.64	1.00	0.10	0.016	0.070	0.228	1.006
	SG4	4.05	1.80	0.71	0.20	0.020	0.098	0.175	0.726
	SG5	4.20	2.14	0.78	0.11	0.025	0.096	0.234	0.596
	SG6	2.86	2.27	0.84	0.19	0.005	0.099	0.334	0.904
Beam	SG3-B1	4.19	2.64	1.00	0.10	0.013	0.067	0.192	0.797
	SG4-B1	4.05	1.80	0.71	0.20	0.027	0.088	0.209	0.711
	SG5-B1	4.20	2.14	0.78	0.11	0.026	0.071	0.215	0.476
	SG6-B1	2.86	2.27	0.84	0.19	9.5×10^{-9}	0.084	0.240	0.729
Wedge	SG3-W1	4.19	2.64	1.00	0.10	0.020	0.069	0.243	1.034
	SG4-W1	4.05	1.80	0.71	0.20	0.007	0.107	0.144	0.567
	SG5-W1	4.20	2.14	0.78	0.11	0.025	0.124	0.231	0.584
	SG6-W1	2.86	2.27	0.84	0.19	0.018	0.110	0.399	0.904

6.3.2 Effect of ligament length on softening curve

In order to examine the effect of the specimen size on the tailing part softening curve, relative variation of crack opening parameters to the averaged softening curve was analyzed. The relative variation of the crack opening parameters is expressed by the following equation.

$$r_i = s_i v_i = \frac{w_{ia,SG1} + w_{ia,SG2} + w_{ia,LG} + w_{ia,WG}}{w_{ca,SG1} + w_{ca,SG2} + w_{ca,LG} + w_{ca,WG}} v_i, \quad v_i = \frac{w_i}{w_{ia}} \qquad (6-1)$$

In the equation, r_i is the relative variation of the crack opening parameter, s_i is a scale factor for the whole test series and represents the relative magnitude of the ith COD parameter to the COD at zero cohesive stress, and v_i is the variation ratio of the ith COD parameter (w_i) to the ith COD parameter of the averaged softening curve (w_{ia}). The COD parameter w_i for each specimen is listed in Table 6-2, and w_{ia} is the COD parameter of the averaged softening curve for the different size specimens, which are listed in Table 6-1. The relative variation of the crack opening parameter r_i for the beam and wedge splitting specimens of different size is plotted over their ligament length, as shown in Fig. 10. In Fig. 10, the 45° line is the regression line for each r_i. In both the beam and wedge splitting specimens, r_c corresponding to w_c increases with an increase of the ligament length, and the effect of the specimen size or the ligament length on the COD parameter is reduced for the relatively smaller COD parameter. From Figs. 6-8~6-10, it can be seen that the initial branch of the softening curve is not greatly affected by the size, and the tail part of the curve becomes longer with an increase of size.

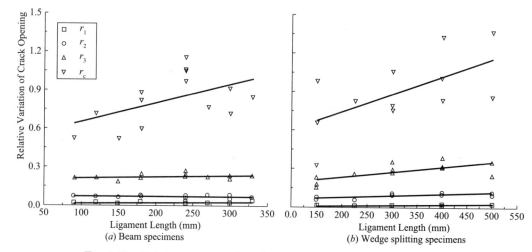

Fig. 6-10   Relative variation of crack opening according to ligament length

6.3.3   Comparison between softening curves of beam and wedge splitting tests

The test series of SG3 to SG6 in the preceding paper were designed for a comparison between the different geometries of beam and wedge splitting specimens. Their ligament length is slightly different: 150mm for the beam specimens and 180mm for the wedge splitting specimens. Although there is a size effect on the softening curve, a 30mm difference is not expected to substantially affect the softening curve, and would not influence a comparison between softening curves of the beam and wedge splitting specimens. For the comparison, the variation ratio v_i defined above was calculated from Table 6-3 and is shown in Fig. 6-11. Variation of the parameters w_1, w_2 and w_3 between the beam and wedge specimens is not large and there is no specific trend according to the different geometry. In contrast with parameters, w_c of the wedge specimen is much larger than that of the beam specimen. Similar to the size effect of the softening curve, the parameter w_c is largely affected by the geometry.

6.3.4   Possible mechanism for the size and geometry effect of fracture energy

The correlation of the softening curve to the mechanism of FPZ has been previously studied[12] and is schematically described in Fig. 7-12 (a). Microcracks start to form immediately before the stress reaches the tensile strength. After the tensile strength is reached, the microcracks are localized and extended in the potential line of a macro visible crack. The first steep branch of the softening curve corresponds to localization and extension of the microcracks. According to the features of the softening curve found above, the process until this fracture stage is not substantially affected by the specimen size or geometry. The stress transmitted to the region where microcracks are distributed by the bridging effect remains even after separation of the crack surface. The region corresponds to the tail of the softening curve and is the origin of the observed size-dependent fracture energy.

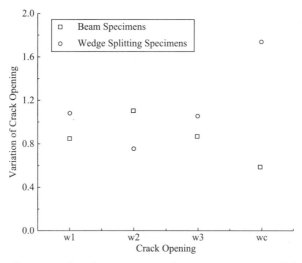

Fig. 6-11 Variation of crack opening according to specimen type (SG3 to SG6)

As shown above, the tail part becomes longer with an increase of specimen size. In other words, the fracture energy dissipated in the region to which the stress is transmitted by the bridging effect is larger for a larger size specimen. The energy is dissipated in the site of microcracking. If the density of the damage due to the microcracks is uniform according to the size, the area that the bridging affects or the width of FPZ should be larger for the larger size specimen, as shown in Fig. 6-12(*b*).

According to previous experimental and analytical studies[12-14], the FPZ is not symmetric and its width varies along the visible crack line. In Hu and Wittmann's[15] work and Hu and Duan's[16] work, the variation of the width along the crack line and the shape of FPZ according to the size result in the observed size-dependent fracture energy. This was analyzed by introducing a local fracture energy concept. Their findings are partly supported by the results of this study.

Although the size effect of the fracture energy was found within the range of specimen size tested in the preceding paper, it is clear that the fracture energy does not continue increasing over the size, and asymptotically approaches a certain value. The width of the FPZ becomes narrow as the crack propagates close to the back boundary of the specimen, which is the end point of the ligament[13,16]. The influence of the boundary on the measured fracture energy might be a major cause of the size-dependent fracture energy and distinctly reduces with an increase of the size. However, a more precise analytical study needs to be carried out in order to fully understand the effect of size and geometry on the softening curve found in this study as well as the size effect and asymptotic feature of the fracture energy.

6 Effect of Specimen Size on Fracture Energy and Softening Curve of Concrete: Part Ⅱ. Inverse Analysis and Softening Curve

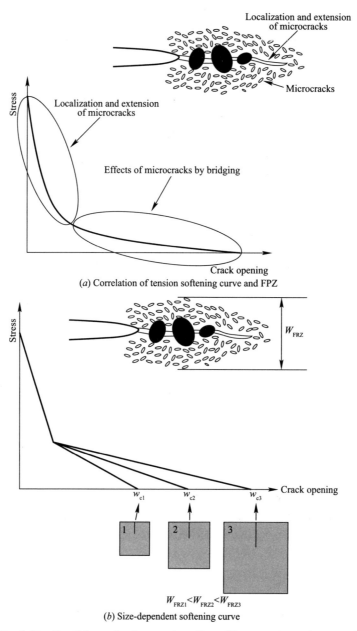

Fig. 6-12 Possible mechanism on size effect of fracture energy (continue)

6.4 Conclusions

The softening curve was assumed as a tetra-linear curve and found from an inverse analysis for every specimen tested in the preceding paper, and the effect of specimen size and geometry on the softening curve was investigated. The first steep branch of the softening curve is similar for specimens of different size but the tail of the curve becomes longer with an increase of the specimen size. The experimentally observed size-dependent fracture energy can be explained by the observed features on the softening curve. Finally, a possible mechanism of the fracture process zone with respect to the size effect and asymptotic feature was discussed. In the future, a more precise analytical study needs to be carried out in order to fully understand the effect of size and geometry on the softening curve as well as the size effect and asymptotic feature of the fracture energy.

References

[1] Z.P. Bazant, B.H. Oh, Crack band theory for fracture of concrete, Materials and Structures 16 (1983) 155-177.

[2] A. Hillerborg, M. Modeer, P.E. Petersson, Analysis of crack formation and crack growth in concrete by means of fracture mechanics and finite elements, Cement and Concrete Research 6 (6) (1976) 773-782.

[3] G. Cusatis, Z.P. Bazant, L. Cedolin, Confinement-shear lattice model for concrete damage in tension and compression: I. theory, ASCE Journal of Engineering Mechanics 129 (12) (2003) 1439-1448.

[4] G. Cusatis, Z.P. Bazant, L. Cedolin, Confinement-shear lattice model for concrete damage in tension and compression: II. computation and validation, ASCE Journal of Engineering Mechanics 129 (12) (2003) 1449-1458.

[5] Z.P. Bazant, G. Pijaudier-Cabot, Nonlocal continuum damage localization instability and convergence, ASME Journal of Applied Mechanics 55 (1988) 287-293.

[6] V.S. Gopalaratnam and B.S. Ye, Numerical characterization of the nonlinear fracture process in concrete, Engineering Fracture Mechanics, 40(6).

[7] Z. Zhao, S.H. Kwon, S.P. Shah, Effect of specimens size on fracture energy and softening curve, Cement and Concrete Research (2007) Submitted with the this manuscript.

[8] J.K. Kim, Y. Lee, S.T. Yi, Fracture characteristics of concrete at early ages, Cement and Concrete Research 34 (2004) 507-519.

[9] K.M. Brown, Derivative-free analogues of the levenberg, Marquardt and Gauss Algorithms for Non-linear Least Square Approximations. IBM, Philadelphia Scientific Center Technical Report, No. 320-2994, 1970.

[10] S.H. Kwon and S.P. Shah, Model to Predict Early-Age Cracking of Fiber-Reinforced Concrete due to Restrained Shrinkage, ACI Material Journal, Submitted for publication.

[11] Z.P. Bazant, J. Planas, Fracture and Size Effect in Concrete and Other Quasibrittle Materials, CRC Press, New York, 1998.

[12] N. Nomura, H. Mihashi, M. Izumi, Correlation of fracture process zone and tension softening behavior in concrete, Cement and Concrete Research 21 (1991) 545-550.

[13] J.P.B. Leite, V. Slowik, H. Mihashi, Computer simulation of fracture processes of concrete using mesolevel models of lattice structures, Cement and Concrete Research 34 (2004) 1025-1033.

[14] L. Cedolin, S.D. Poli, I. Iori, Experimental determination of the fracture process zone in concrete, Cement and Concrete Research 13 (1983) 557-567.

[15] X.Z. Hu, F.H. Wittmann, Fracture energy and fracture process zone, Materials and Structures 25 (1992) 319-326.

[16] X.Z. Hu, K. Duan, Influence of fracture process zone height on fracture energy of concrete, Cement and Concrete Research 34 (2004) 1321-1330.

7 Influence of Coarse Aggregate Size on Softening Curve of Concrete

7.1 Introduction

The softening curve is closely related to the concrete fracture behavior. The softening curve parameters are basic parameters of the fracture behavior of concrete.

In recent years, researchers started to adopt inverse analysis method based on simple fracture tests to study the softening curve of concrete material. The method is combined with fracture tests and adopts the finite element method based on concrete fracture mechanics model to make numerical simulation for obtaining the concrete softening curve. Compared with concrete direct tension test, three-point bending notched beam test and wedge splitting test are simple fracture tests, which can be conducted in general labs. Especially the wedge splitting test, which is of small weight-volume ratio and relatively long ligament, can avoid the influence of the specimen weight on the experimental results. Meanwhile the wedge type loading method reduces the requirements of stiffness to the testing machine to simplify the test operation.

The paper, based on the load-crack mouth opening displacement (*P-CMOD*) curve of wedge splitting test, compiled an inverse analysis program to determine the concrete softening curve. Four-linear softening curves were obtained by the program. The finite element mesh generation method for the inverse analysis program was studied, as well as the method to determine the initial parameters of softening curve. The influence law of coarse aggregate size on the softening curves was analyzed.

7.2 Experiments

The test data are selected from the fracture tests[1] conducted by our research group. Totally 3 groups of concrete wedge splitting specimens with different coarse aggregate size d_{max} are chosen as is shown in Table 7-1. The specimens are made under the same mix proportion which sizes are all 200mm×300mm×300mm. After the specimen molding and maintaining for a certain age, the wedge splitting tests were performed to obtain the *P-CMOD* curves.

Material properties of wedge splitting specimens Table 7-1

Specimens	d_{max} (mm)	E (GPa)	γ	Number of companion specimens
WS1	10	31.4	0.171	3
WS2	20	39.3	0.202	3
WS3	40	37.1	0.190	3

7.3 Inverse Analysis of Softening Curves of Concrete

In order to find the softening curve that optimally fits the measured *P-CMOD* curve, a finite element analysis and an optimization technique are incorporated. The algorithm of the inverse analysis is shown in Fig. 7-1. The initial tangent modulus E and poisson ratio γ of the concrete specimens, listed in Table 7-1, are used as material input for the specimens. The cohesive crack model[2] and the principle of superposition suggested by Gopalaratnam and Ye[3] are employed to simulate the crack propagation. The overall optimization procedure is based on the Levenberg-Marquardt[4] method.

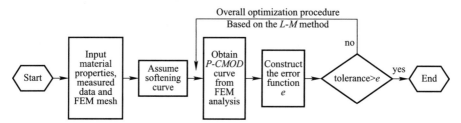

Fig. 7-1 Algorithm of inverse analysis

7.4 Procedure of Inverse Analysis

Fig. 7-2 shows the procedure of the inverse analysis to find the four-linear softening curve optimally fitting the measured *P-CMOD* curve for each specimen. Two test data sets were prepared prior to the inverse analysis, the test data extraction (from the original test data of *P-CMOD* curve) method was described in paper[5]. The first test data set of a *P-CMOD* curve consists of 20 points (8 points in the ascending part, 12 points in the descending part). The last *CMOD* value of the first set is almost half of the largest *CMOD* measured in the test. In the second set, which also consists of 20 points, the ascending part is the same as in the first set, and the descending part is roughly twice longer than that of the first set.

7.5 Determination of Initial Softening Parameters

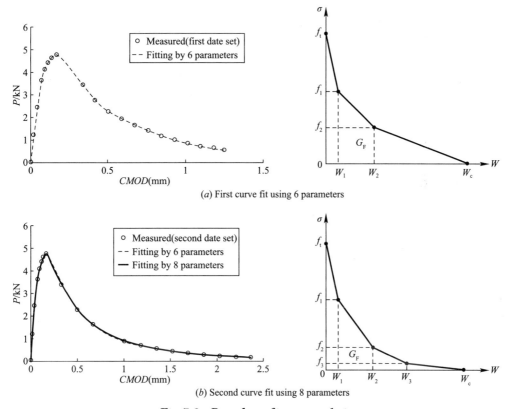

(a) First curve fit using 6 parameters

(b) Second curve fit using 8 parameters

Fig. 7-2 Procedure of inverse analysis

In the first step of the inverse analysis, the three-linear softening curve optimally fitting the first data set was found, as shown in Fig. 7-2(a), and six parameters of the three-linear softening curve were obtained. In the second step, three parameters obtained from the first curve fit, f_t, f_1, and w_1, which do not affect the tail part of the P-CMOD curve, were fixed and the remaining five parameters among the eight parameters of the four-linear curve fitting the second data set were found, as shown in Fig. 7-2(b). As shown in the left graph of Fig. 7-2(b), the four-linear softening curve can more accurately fit the tail part of the P-CMOD curve.

7.5 Determination of Initial Softening Parameters

As shown in Fig. 3, to completely determine the shape of three-linear softening curve needs to determine the following 6 parameters: the absolute value of slope of the first straight line $k_1(\tan\theta_1)$ as well as the stress value f_1 at the corresponding turning point 1, the absolute value of slope of the second segment k_2 ($\tan\theta_2$) as well as the stress value f_2 at the corresponding turning point 2, the tensile strength f_t and the maximum crack width w_c.

According to the observation results of concrete direct tension test, parameters among the six can be preliminary assumed that the f_t, w_c are respectively equivalent to those obtained by the direct tension test. This is just the first "guess" of inverse analysis calculation.

7 Influence of Coarse Aggregate Size on Softening Curve of Concrete

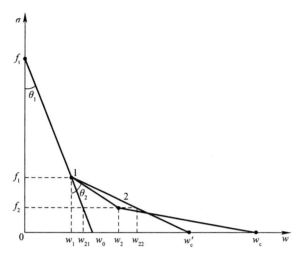

Fig. 7-3 The schematic diagram of initial setting of a three-linear softening curve

As the stiffness of the decline segment of softening curve should be gradually reduced, so $\tan\theta_1$ should be less than $\tan\theta_2$. At the same time compare it to the bilinear softening curve, as the fracture energy is equal, thus it is obtained that the area under the three-linear softening curve is equal to the area under the bilinear softening curve according to being equal to in fracture energy, so $\tan\theta_2$ should be less than the absolute value of the slope for the second straight line of the corresponding bilinear softening curve. According to the above analysis, it is known that the crack width w_2 at turning point 2 in Fig. 7-3 should be between w_{21} and w_{22}. Similarly, assume that the stress value $f_1=\eta f_t$ at turning point 1, the stress value $f_2=\alpha f_1 = \alpha\eta f_t$ at turning point 2, the crack width $w_2=\beta(w_{21}+w_{22})$ at turning point 2, thus: $w_1=f_t(1-\eta)k_1$. According to the definition of the fracture energy G_F, the following equation is obtained:

$$\frac{1}{2}(f_1+f_t)w_1 + \frac{1}{2}(f_1+f_2)(w_2-w_1) + \frac{1}{2}f_2(w_c-w_2) = G_F \qquad (7\text{-}1)$$

Where, $k_1 = \dfrac{2K_{IC}^2}{Ef_t^2}$, $w_{21} = w_c - \alpha(w_c - w_1)$, $w_{22} = w_c' - \alpha(w_c' - w_1)$, $w_c' = \dfrac{1}{\eta}\left(\dfrac{2G_F}{f_t} - w_1\right)$.

The fracture energy G_F, fracture toughness K_{IC} can be obtained by direct tension test and wedge splitting test of the same material. Thus, the positions of turning points 1, 2 can be determined by the three parameters, namely, η, α and β. Reference[6] suggests that the three parameters can be selected as $\eta=0.3$, $\alpha=0.25$ and $\beta=0.5$ for the concrete at age of 28 days.

Based on the above approach, 6 parameters of the three-linear softening curve can be preliminary determined, to complete the "first guess" of the inverse analysis calculation. Similarly, the initial parameters of the four-linear softening curve can also be obtained.

7.6 FEM Mesh Generation for the Inverse Analysis Calculation

For wedge splitting specimen, considering the symmetry, half of the specimen is taken for building FEM model (as shown in Fig. 7-4). The calculation was based on the plane stress problem. Smaller the ligament mesh size is, the closer $P\text{-}CMOD$ curve calculated to the measured one, the softening curve solved by inverse analysis is more precise also. But if mesh size is too small, the mesh point would be too fine, which will reduce the efficiency of inverse analysis calculation.

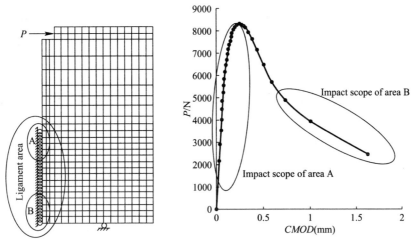

Fig. 7-4 The FEM mesh generation of a wedge-splitting specimen and its influence on the calculated *P-CMOD* curve

As shown in Fig. 7-4, the calculation results obtained by different FEM mesh generation show that: the finer the mesh in ligament A is, the closer of the ascending part of the calculated *P-CMOD* curve to that of the tested one. If the number of nodes in the range of ligament area B is too small, that can not reach the length of the experimental *P-CMOD* curve tail, will lead to solve abnormal softening parameters.

To sum up, different parts of the calculation model adopt meshes with different shapes and sizes. This can not only ensure the calculating precision, and also enhance the inverse analysis calculation efficiency. Finally adopted meshes are as shown in Fig. 7-5.

7.7 Results of Inverse Analysis Calculation

Table 7-2 shows the obtained 8 parameters of four-linear softening curve for each wedge-splitting specimen group with different coarse aggregate size d_{max}.

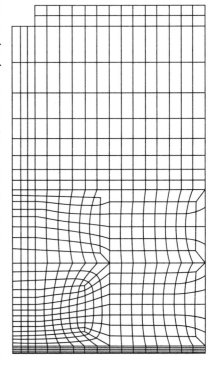

Fig. 7-5 The finally adopted FEM mesh generation for inverse analysis calculation

7 Influence of Coarse Aggregate Size on Softening Curve of Concrete

Parameters of four-linear softening curve for each specimen group   Table 7-2

Specimens	f_t (MPa)	f_1 (MPa)	f_2 (MPa)	f_3 (MPa)	w_1 (mm)	w_2 (mm)	w_3 (mm)	w_c (mm)
WS1	3.39	1.58	0.418	0.100	0.004	0.132	0.350	0.700
WS2	3.36	1.85	0.427	0.180	0.0244	0.135	0.280	0.860
WS3	3.75	2.40	0.972	0.120	0.00642	0.102	0.330	0.904

7.8 Influence of Maximum Aggregate Size on Softening Curve of Concrete

Tensile strength f_t, maximum crack width w_c and fracture energy G_F are three control parameters of softening curve of concrete.

As shown in the Fig. 7-6, all the control parameters which are f_t, w_c and G_F, increase with the increase of d_{max}. Moreover, the increase amplitudes of the w_c and G_F become smaller.

The study of Wu and Zhao[7] stated that G_F reaches its maximum value when d_{max}=40mm. Furthermore, they gave a correction coefficient table of G_F which contains the correction coefficient of G_F based on that of d_{max}=20mm (as listed in Table 7-3). As shown in the Fig. 7-6, the ratio of G_F of d_{max}=40mm to that of d_{max}=20mm is 1.19. This ratio coincides with the correction coefficient 1.22 that provided in the Table 7-3.

Correction coefficient of G_F versus d_{max} according to the paper[7]   Table 7-3

d_{max} (mm)	20	40	80	150
Correction coefficient of G_F	1.0	1.22	1.11	1.14

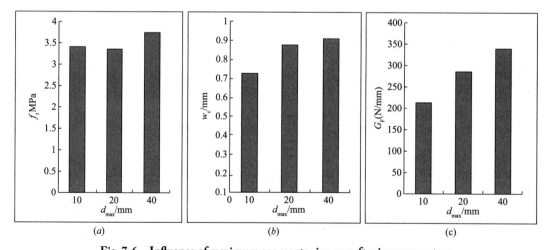

Fig. 7-6 Influence of maximum aggregate size on softening parameters

7.9 Conclusions

Based on the wedge splitting test results and Fictitious Cracking Model, four-linear softening curves of three groups specimens with different coarse aggregate size d_{max} were obtained. The influence of FEM mesh generation on the result of inverse analysis was investigated, a FEM mesh generation method which refine the both ends of ligament was proposed. The results of inverse analysis show that all the control parameters of softening curve, which are the tensile strength f_t, maximum crack width w_c and fracture energy G_F, increase with the increase of d_{max}. Moreover, the increase amplitudes of the w_c and G_F become smaller.

References

[1] Hougui Zhou, Qingbin Li and Zhifang Zhao. Research on fracture simulation of mass concrete cracks, Cooperative Research Report of China Gezhouba(Group) Corporation, Tsinghua University and Yantai University (2004) (In Chinese)

[2] A. Hillerborg, M. Modeer, P. E. Petersson: Cement and Concrete Research. Vol. 6 (1976), p. 773-82

[3] V.S. Gopalaratnam and B.S. Ye: Engineering Fracture Mechanics. Vol. 40 (1991), p.991-1006

[4] K.M.Brown: Numerische Mathematik Vol. 18(1972), p. 289-297

[5] Zhifang Zhao, Lijian Yang, Zhigang Zhao and Minmin Zhu (2009). Test Data Processing Method of Fracture Experiments of Dam Concrete for Inverse Analysis, Proceedings of the International Symposium on Computational Structural Engineering, June 22-24, 2009, Tongji University, Shanghai. p. 1239-1247

[6] Jian Shen. Numerical Analysis of Softening Relationship for Early-Age Concrete, Master's Dissertation of Zhejiang University (2004) (In Chinese)

[7] Zhimin Wu and Guofan Zhao: Journal of Dalian University of Technology Vol.34 (1994), p. 583-588 (In Chinese)

8 Research on Softening Curve of Concrete and Fracture Energy by Different Methods

8.1 Introduction

The fictitious crack model[1] is typical approach to simulate fracture process of cement-based materials in structures. If fracture process of civil and hydraulic engineering structures are simulated by employing the models, the experimentally determined fracture parameters, such as tensile strength f_t, fracture energy G_F and softening curve (σ-w curve) have to be needed.

The existing experimental and theoretical research shows that the softening relationships (σ-w curves) of cement-based materials control static and fatigue crack propagation in structures so that it have an effect on strength of the structures. Macroscopical fracture behaviors and strength of structures made of various concrete materials could be predicted by computer modeling and estimating mechanical properties of concrete materials which have different components (the estimation gives priority to σ-w curves). It can provide scientific foundation for materials design based on structural properties.

Researchers developed in succession several representative determination method of σ-w curve of concrete materials as follows: (1) Direct tension method.[2] However, direct tension test has strict requirements for testing machine and has to deal with the loading concentricity. Thus, it is difficult to succeed. In addition, the test is expensive and time consuming. The additional problems are induced by the effect of strain gradients in the ligament due to non-uniform cracking. Direct tension test is less applicable to the large size dam concrete specimens. (2) Researchers begin to find a new method to identify σ-w curve because of defects of direct tension test . Li[3] proposed a method to experimentally determine σ-w curve based on J-integral principal. The σ-w curve can be obtained by experimentally determining the load-loading point displacement (P-δ curve) of two specimens. However, a great error will be induced between the two curves by any error of the measured curve, which can give rise to distortion of the calculated σ-w curve.

In recent years, researchers identify the complete softening curve of concrete materials by employing inverse analysis method. This method based on fracture experiments of concrete is numerical simulations by using discrete approach of non-linear fracture mechanics. First, a softening curve is assumed and the load-displacement curve (such as load-crack mouth opening displacement P-$CMOD$ curve) obtained by the numerical simulation is compared to the one determined experimentally by the fracture experiment of concrete. By updating the assumed softening curve, the numerical results optimally fit the measured ones. When the best fit of the numerical and the experimental results is obtained, the assumed softening curve is viewed as the one for characterization of the material behavior.

In this chapter, a calculating program for inverse analysis based on Levenberg-Marquardt (L-M) optimization algorithm by Fortran language is developed and the concrete softening curves are obtained by inverse analysis. Comparison and analysis of concrete fracture energy obtained by inverse analysis method and by work of fracture method recommended by RILEM is carried out and size effect of concrete fracture energy is discussed.

8.2 Experiment and Data Processing

8.2.1 Experiment

Diagram of standard three-point bending notched beams is shown in Fig.8-1 and the size of specimens with different height is shown in Table 8-1. The beam thickness is 120mm and the notch width is 2mm. Table 8-2 shows the concrete mix proportion. Stable P-$CMOD$ and Load-displacement (P-δ) curves were obtained when three-point bending beam was in displacement-controlled loading. The cubic compressive strength of concrete is 43.3 Mpa and test set-up for TPB is shown in Fig. 8-2, more details about the test see chapter 5 and reference[4].

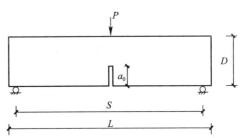

Fig. 8-1 Diagram of Three-point bending notched beams

Size of specimens　　　　　　　　Table 8-1

Specimens	Dimension of specimen (mm)			
	D	S	L	a_0
TPB-1	100	400	500	40
TPB-2	150	600	700	60
TPB-3	200	800	900	80
TPB-4	300	1200	1300	120
TPB-5	400	1600	1700	160
TPB-6	500	2000	2100	200

Mix propotion of concrete　　　　　　　　Table 8-2

Maximum gravel size (mm)	W/B	Unit weight (kg/m³)						
		C	F.A	W	S	G	W.R.	A.E.
20	0.50	185	79	132	846	1152	1.584	0.0185

8.2.2 Data Processing

Inverse analysis, based on the assumption that the result of experiment is the true response of structure, needs accurate experimental data to determine concrete softening curves. Because typical experimental data always contain measurement errors (outliers and noise) caused by environment, equipment and human errors, meanwhile, for improving calculating efficiency, it is necessary to process data. There were 3-4 effective specimens for each group standard three-point bending notched beam, it is necessary to achieve *P-CMOD* curve representing for each group specimen for inverse analysis. The original test data were processed by the data processing method [5] and the results are shown in Fig. 8-3.

Fig. 8-2 Test set-up for TPB test

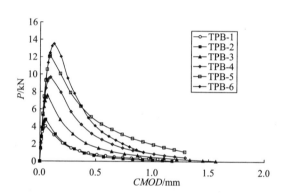

Fig. 8-3 Tested *P-CMOD* curves for each group specimen

8.3 Levenberg-Marquardt Optimization Algorithm[6-7]

The Levenberg-Marquardt method is an iterative algorithm that locates the minimum of a multivariate function and can be seen as an interpolation between gradient descent and Gauss-Newton. It has become a standard technique for non-linear least squares problems. When the current solution is far from the correct one, the algorithm behaves like a steepest descent method: slow, but guaranteed to converge. When the current solution is close to the correct solution, it becomes a Gauss-Newton method.

Optimization is equivalent to the determination of minima of a cost function, which is given as

$$E = \sqrt{\sum_{1}^{n}(P_{\exp} - P_{\text{cal}})^2} \tag{8-1}$$

When the cost function comes to the minimum, the assumed softening curves is the true value.

8.4 Inverse Analysis and Results

When FCM and FEM was employed to simulate crack propagation, all the nodes along the ligament were treated as follows: when the stress of the node reaches the tensile strength, the node was released and the principle of transfer for the cohesive force suggested by Gopalaratnam and Ye[8] was employed. Specimen self-weight is incorporated. The assumption that all of the structure except for the fracture process zone is elastic simplifies the calculation. Young's modulus E and Poisson's ratio v for this plane stress problem were obtained from tests. Considering its symmetry, only half of the specimen was modeled, in this way, the number of degrees of freedom is reduced significantly as compared to a two-dimension full model and the computational efficiency can be improved. Fig. 8-4 shows the finite element meshes for the beam.

Petersson[9] proposed a two-branch (bilinear) softening law with a fixed breakpoint, however, Petersson's two-branch law may not correctly predict the softening behavior. In this study, tri-linear softening curve (Fig. 8-5) is adopted to simulate the fracture process of three-point bending beam. Table 8-3 shows the six parameters of the tri-linear softening curve for each specimen. Fig. 8-6 shows the comparison of the experimental and calculated P-$CMOD$ curves for each group specimens, it can be seen that those results fit very well.

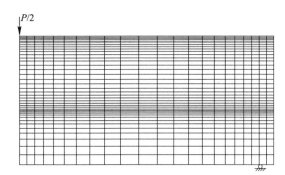
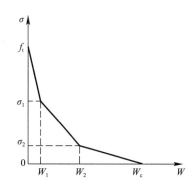

Fig. 8-4 FEM meshes   **Fig. 8-5 Tri-linear softening curve**

Parameters of tri-linear softening curve for each specimen   **Table 8-3**

Specimens	Parameters of tri-linear softening curve					
	f_t(MPa)	σ_1(MPa)	σ_2(MPa)	w_1(mm)	w_2(mm)	w_c(mm)
TPB-1	3.23	1.36	0.44	0.0351	0.0931	0.358
TPB-2	3.54	1.09	0.30	0.0490	0.161	0.426
TPB-3	3.53	2.14	0.44	0.0243	0.0783	0.566
TPB-4	3.22	1.98	0.55	0.0258	0.0941	0.472
TPB-5	3.30	1.23	0.42	0.0399	0.187	0.764
TPB-6	3.28	1.66	0.41	0.0294	0.113	0.489

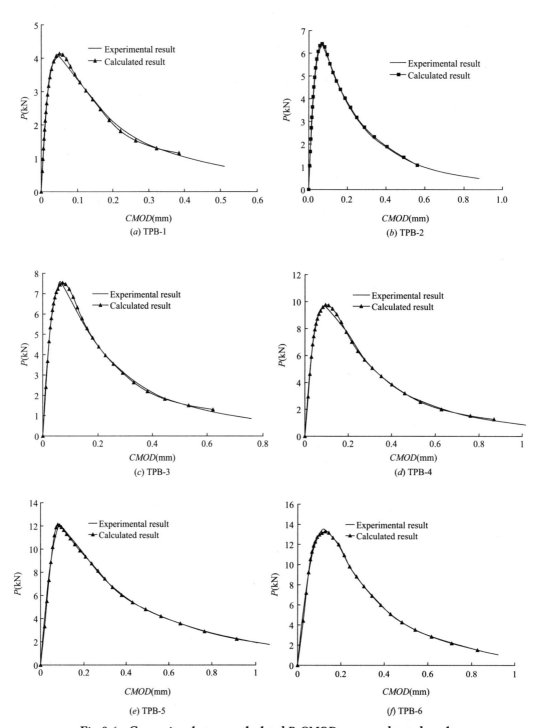

Fig. 8-6 Comparison between calculated *P-CMOD* curves and tested results

8.5 Comparison and Analysis of Concrete Fracture Energy by Two Methods

8.5.1 The 1st method recommended by RILEM to determine concrete fracture energy

Fracture energy of concrete, which is the energy required to open unit area of crack surface, is an important parameter for concrete fracture mechanics. A three-point bending beam was recommended by RILEM to estimate the concrete fracture energy,

$$G_F = \frac{W + mg\delta}{A} \quad (8\text{-}2)$$

where, W—the area under Load-displacement (P-δ) curve;

m—mass of the specimen;

g—gravity;

δ—ultimate displacement of beam;

A—ligament area.

8.5.2 The 2nd method based on softening curve to determine concrete fracture energy

Base on FCM, the value of concrete fracture energy is the area under softening curve σ-w and can be obtained by the following expression:

$$G_F = \int_0^{w_c} \sigma(w) \mathrm{d}w \quad (8\text{-}3)$$

Where, w_c is the maximum crack width when stress is zero in the softening curve.

8.5.3 Comparison of concrete fracture energy by the two methods

The results of the concrete fracture energy for each group specimens by the above two methods is shown in Fig.8-7, and it can be seen that the value of concrete fracture energy varies according to beam height for both of the above methods and show the same trend, namely, fracture energy increases with growth of specimen size when the height of beams varies for 100mm to 400mm, and when the height of beams reached 500mm, fracture energy decreased. Studies[10-12] have demonstrated that concrete fracture energy didn't increase any more when the specimen size reach to a certain

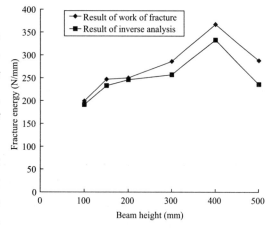

Fig. 8-7 Variation of fracture energy via beam height

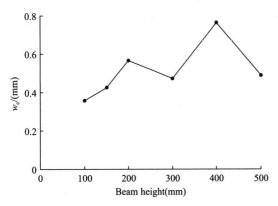

Fig. 8-8 Variation of w_c via beam height

value.

For the limitation of the specimens, the result of those specimens with height above 500mm was not available, further study needs to be carried out.

The w_c via beam height was shown in Fig. 8-8. It can be seen that w_c has the trend that it increases with growth of specimen size and tends to increase no longer when the specimen size attains to a certain level. Energy dissipation within FPZ increases with growth of specimen size. If the energy dissipation is uniform along the ligament, the area that the bridging affects or the width of FPZ increases with the growth of specimen size. In Hu's [13] work, boundary effect is the main cause for size effect of concrete fracture energy and the size effect becomes less with the increase of specimen size.

8.6 Conclusions

Concrete softening curves of six group three-point bending beams with different dimension are obtained by inverse analysis. Concrete fracture energy is obtained by inverse analysis method and work of fracture method recommended by RILEM, meanwhile, and size effect of concrete fracture energy is discussed. The results show as follows: (1) the fracture energy by the two methods show the same trend, namely, fracture energy increases with growth of specimen size and tends to increase no longer when the specimen size attains to a certain level. (2) the w_c increases with growth of specimen size and tends to increase no longer when the specimen size attains to a certain level, size effect of softening curve may explain the size effect of fracture energy.

References

[1] Hillerborg. M., M. Modeer, and P. Peterson. Analysis of crack formation and crack growth in concrete by means of fracture mechanics and finite elements [J]. Cement and Concrete Research, 1976, 6: 773-782.

[2] Gopalaratnam V S, Shah S P. Softening response of plain concrete in direct tension [J]. ACI Journal, 1985, 82(3): 310-323.

[3] Li, V.C., Chan, C-M and Lung, C.K.Y. .Experimental determination of the tension-softening relations for cementitous composites. Cement and Concrete Research, 1987, 17(3):441-452

[4] ZHAO Zhi-fang. Research on fracture behavior of dam concrete based on the crack cohesive force[D]. Postdoctoral Research Report of Tsinghua University and China Gezhouba Group Co., Beijing:Tsinghua University, 2004 (in Chinese)

References

[5] Zhifang Zhao, Lijian Yang, Zhigang Zhao and Minmin Zhu. Test Data Processing Method of Fracture Experiments of Dam Concrete for Inverse Analysis[C]. Proceedings of the International Symposium on Computational Structural Engineering, Shanghai, China, June 22-24, 2009. 1239-1247

[6] LIU Zhi-ming, WANG Zhong-xian, SU Hong et al. Prediction of stress concentration based on Levenberg-Marquardt algorithm [J].Journal of Jiangsu university of science and technology (natural science). 2001, 22(6): 84-87 (in Chinese)

[7] LI Gui-ling, WAN Jian-hua, TAO Hua-xue. Nonlinear data processing based on improved marquardt method. [J]. Journal of Institute of Surveying and Mapping, 2001, 18(13):167-169 (in Chinese)

[8] V.S. Gopalaratnam and B.S. Ye. Numerical characterization of the nonlinear fracture process in concrete [J]. Engineering Fracture Mechanics, 1991, 40(6): 991-1006

[9] Petersson P. E. Crack growth and development of fracture zones in plain concrete and similar materials[D]. Lund Institute of Technology, Lund, Sweden, 1981

[10] WU Zhi-min, ZHAO Guo-fan. Size effect of fracture energy of concrete [J]. Industrial Architecture, 1995, 25(11): 25-28 (in Chinese)

[11] Bazant Z P, Planas J. Fracture and size effect in concrete and other quasibrittle materials [M]. Boca Raton: CRC Press, 1998, pp 275-289

[12] Wittmann F. H, Mihashi H. and Normura N. Size effect on fracture energy of concrete [J]. Engineering Fracture Mechanics, 1990, 35: 107-115

[13] Kai Duan, Xiaozhi Hu, Folker H. Wittmann. Size effect on specific fracture energy of concrete [J]. Engineering Fracture Mechanics, 2007, 74: 87-96

9 Research on Softening Properties of Concrete Based on Three-point Bend Beam Tests

9.1 Introduction

The softening behavior is one of the important fracture characteristics of concrete, which can be characterized by three parameters, tensile strength f_t, maximum cohesive crack width w_c and fracture energy G_F. G_F is an area under a softening curve (σ-w curve) of concrete.

There were some studies on fracture energy of concrete[1-6]. However, comprehensive and systematic research on softening properties of concrete is still scarce.

The stable fracture experiments of three-point bend beams (TPB) which are 12 groups and different sizes were performed in this research. The inverse analysis procedure was programmed by employing the Fictitious Crack Model and Levenberg-Marquardt optimization algorithm. Based on the experimental results and program, the softening curves of concrete were achieved by finding the softening curve that optimally fits the tested load-crack mouth opening displacement (P-CMOD) curve. The effects of aggregate size, specimen size and concrete strength on softening properties were also investigated.

9.2 Three-point Bend Notched Beam Tests

9.2.1 Materials and mixes

As shown in Fig. 9-1, a_0 is the length of pre-crack, B and D are the thickness and height of specimen, respectively. The width of pre-crack is 2 millimetres. There are 12 groups beams, which include three specimens for each group. The specimen dimensions are showed in Table 9-1, and the specimen thickness B is 120 millimetres for all specimens.

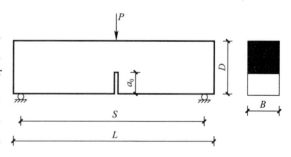

Fig. 9-1　Diagram of three-point bend notched beam

9.2 Three-point Bend Notched Beam Tests

Design grade, aggregate size and dimensions of TPB specimen Table 9-1

Specimen	Design Grade	Aggregate size (mm)	Dimension (mm)			
			D	L	S	a_0
TPB-1	C20	10	300	1300	1200	120
TPB-2			400	1700	1600	160
TPB-3		20	100	500	400	40
TPB-4			150	700	600	60
TPB-5			200	900	800	80
TPB-6			300	1300	1200	120
TPB-7			400	1700	1600	160
TPB-8			500	2100	2000	200
TPB-9	C30		300	1300	1200	120
TPB-10	C40					
TPB-11	C50					
TPB-12	C20	40				

The concrete mix proportions is listed in Table 9-2. All the mixes were designed for a large scale dam built in China. The medium heat cement 525# produced by Jinmen cement plant, I grade fly ash produced by Shandong Zouxian thermal power plant, the superplasticizer JG3 produced by Beijing Yejian, the air-entraining agent DH9 by Shijiazhuang admixture factory were adopted. The fine aggregate and coarse aggregate were crushed sand and crushed gravel, respectively. The mixing water was the local drinking water at the construction site.

Concrete mix proportions Table 9-2

Specimen	W/B	Unit weight (kg/m³)						
		Cement	Fly ash	Water	Sand	Coarse aggregate	Water reducing agent	Air-entraining agent
TPB-1, 2	0.50	196	84	140	869	1090	1.680	0.0196
TPB-3~8	0.50	185	79	132	846	1152	1.584	0.0185
TPB-9	0.45	240	60	135	814	1154	1.800	0.0195
TPB-10	0.35	309	77	135	744	1145	2.316	0.0251
TPB-11	0.30	420	47	140	698	1121	2.802	0.0280
TPB-12	0.50	168	72	120	769	1287	1.440	0.0168

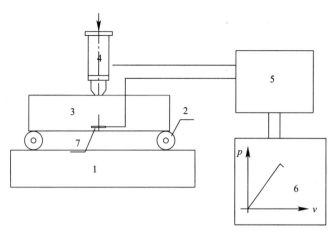

Fig. 9-2　Sketch of test apparatus for TPB specimen

1. Testing machine force bearing platform; 2. roller shaft; 3.specimen; 4. load transducer; 5. DH5937 dynamic strain gauge;
6. automatic acquisition system of computer; 7. displacement transducer

9.2.2　Test procedure

As illustrated in Fig. 9-2 The TPB tests were performed on a very stiff testing machine. A load cell which capacity is 100kN was adopted in the test, and the accuracy was ±2% of the maximum applied load. The crack mouth opening displacement (*CMOD*) was measured by a displacement sensor which capacity and accuracy are 5mm and 0.5μm, respectively. The stable complete Load-CMOD curves of TPB specimens were achieved.

The original test data of *P-CMOD* curves of TPB specimens of dam concrete are very large quantity. The original test data for each specimen reached up to 30000~300000 lines and there were three companion specimens in each specimen group. The representative *P-CMOD* curve which can characterize the fracture property of each specimen group was achieved by filtering, averaging and extraction with the optimization method[7]. Then the obtained *P-CMOD* curve of each specimen group can be as an input for determining the softening curve by inverse analysis method.

9.3　Softening Curve of Concrete Calculated by Inverse Analysis Method

9.3.1　Softening curve determined by FEM based on FCM model

The Fictitious Crack Model (FCM) proposed by Hillerborg can well describe the mechanical properties of the fracture process zone (FPZ)[8]. The principle of superposition suggested by Gopalaratnam[9] was used to simulate the crack propagation. The inverse analysis program for determining softening curve of concrete was developed using Fortran language based on the Levenberg-Marquardt optimization algorithm.

9.3 Softening Curve of Concrete Calculated by Inverse Analysis Method

9.3.2 Softening curve of concrete obtained by inverse analysis

The single linear softening curve, bilinear softening curve, trilinear softening curve and tetralinear softening curve are the common linear softening curves. The more linear sections softening curves has, the more accurate the softening curve is. The schematic diagram of a tetra-linear softening curve is showed in Fig. 9-3.

The inverse analysis program was employed to find the tetra-linear softening curve optimally fitting the tested *P-CMOD* curve for each specimen group. Table 9-3 shows the 8 parameters of the tetra-linear softening curve for every specimen group.

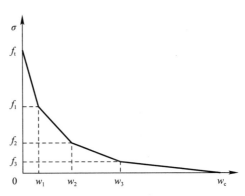

Fig. 9-3 Schematic diagram of a tetralinear softening curve

Tetra-linear softening parameters obtained by inverse analysis Table 9-3

Specimen	Tetra-linear softening parameters							
	f_t	σ_1	σ_2	σ_3	w_1	w_2	w_3	w_c
TPB-1	1.93	0.78	0.29	0.17	0.099	0.245	0.298	0.502
TPB-2	3.10	1.49	0.61	0.21	0.029	0.101	0.302	1.29
TPB-3	3.23	1.36	0.57	0.40	0.035	0.093	0.108	0.388
TPB-4	3.536	1.09	0.73	0.25	0.049	0.072	0.184	0.755
TPB-5	3.53	2.14	0.86	0.22	0.024	0.056	0.183	0.789
TPB-6	3.22	1.98	0.70	0.16	0.026	0.086	0.262	0.838
TPB-7	3.30	1.23	0.19	0.01	0.039	0.166	0.426	1.324
TPB-8	3.27	1.66	0.59	0.24	0.029	0.982	0.210	0.596
TPB-9	3.41	2.34	0.66	0.09	0.024	0.077	0.252	0.650
TPB-10	3.91	2.04	0.94	0.13	0.029	0.056	0.226	0.726
TPB-11	3.29	0.65	0.32	0.10	0.076	0.151	0.282	0.632
TPB-12	2.74	1.77	0.71	0.18	0.038	0.088	0.252	0.938

To verify the rationality of determination of softening curve by inverse analysis, the softening curves obtained by the inverse analysis method of the specimens with the beam height 300mm were compared with those of the corresponding concrete obtained by the direct tension test, as shown in Fig. 9-4.

The results shows that the softening curve by inverse analysis method and that by direct tension test method was almost in good agreement for each specimen. However, the tensile strength of concrete obtained by inverse analysis was larger than that by the direct tension test and the fracture energy obtained by inverse analysis was a little smaller than that by the direct tension test, which was accordance with the results by Uchida and Barr.[10]

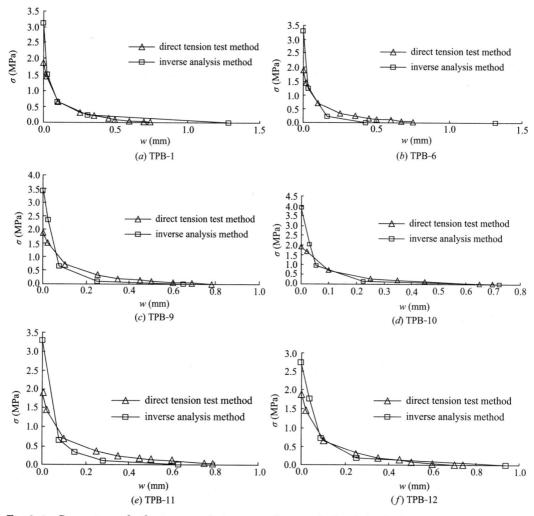

Fig. 9-4 Comparison of softening curve by inverse analysis method and that by direct tension test method

9.4 Softening Behaviors of Concrete

The softening curve of the concrete obtained from the inverse analysis was used for a further study, which is about the influencing rules of aggregate size, specimen size and concrete strength on the softening curve.

9.4.1 Influence of coarse aggregate size of concrete on the softening curve

The TPB specimens with the same size $B \times D \times L = 120\text{mm} \times 300\text{mm} \times 1300\text{mm}$ and different aggregate sizes $d_{max} = 10\text{mm}$, 20mm, 40mm were selected to study.

The influence of maximum aggregate size on fracture energy of concrete is shown in Fig. 9-5(a). The fracture energy increased as the aggregate size increased. Generally, it is believed that the aggregate can hinder propagation of cracks in concrete. It appears that with the increase of aggregate size, the more tortuous the cracks are, and the fracture ligament area as well as fracture energy increases.

The influence of aggregate size on tensile strength of concrete is shown in Fig. 9-5(b). The tensile strength of concrete increased as the aggregate size increased from 10mm to 20mm. When the aggregate size increased from 20mm to 40mm, the tensile strength tends to decrease. It shows no clear trend.

The influence of aggregate size on maximum crack width is shown in Fig. 9-5(c). The maximum crack width of concrete increased as the aggregate size increased. The same trend was observed with the results reported by Mihashi.[11]

The effects of aggregate size, specimen size and concrete strength on the softening curve which was obtained by inverse analysis are further studied.

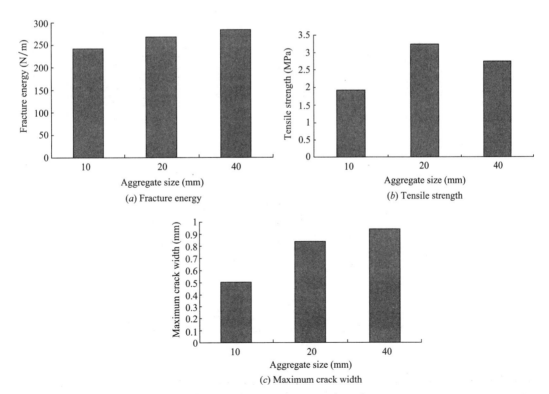

Fig. 9-5 Influence of aggregate size on the softening curve

9.4.2 Influence of specimen size on softening curve of concrete

The specimen TPB3-TPB8 with specimen height varied from 100mm to 500mm were selected for studying the influence of specimen size on softening curve.

The influence of specimen size (beam height) on fracture energy is shown in Fig. 9-6(a).

It appears that the fracture energy increased, then decreased and tends to approach a certain value with the increase of specimen size. This asymptotic behavior was also reported in some other researcher's work.[12-13] The reason maybe related to the mechanism of Fracture Process Zone (FPZ).

The influence of specimen size on tensile strength is shown in Fig. 9-6(b). It was found that the specimen size has little impact on the tensile strength of concrete.

The influence of specimen size on maximum cohesive crack width w_c is shown in Fig. 9-6(c). The trend of the variation of maximum crack width via specimen size is similar to that of fracture energy via specimen size. Namely, it appears the asymptotic behavior. Generally, the larger the specimen size is, the more energy is dissipated in the FPZ. However, Duan's work[14] explained why it exsits asymptotic behavior of fracture energy. It may help to understand the asymptotic behavior of w_c.

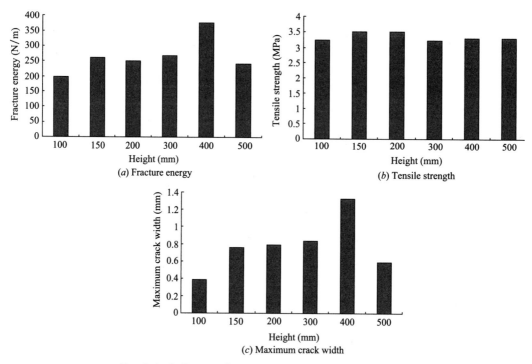

Fig. 9-6 Influence of specimen size on the softening curve

9.4.3 Influence of concrete strength on softening curve

The beam TPB6, TPB9-TPB11 which has the same spciemen size $B \times D \times L$=120mm×

300mm×1300mm and the same coarse aggregate size 20mm were selected for studying effect of concrete strength on softening curve.

The influence of concrete strength on fracture energy is shown in Fig. 9-7(*a*). No clear trend was found for the variation of fracture energy via compressive strength of concrete.

The influence of compressive strength on tensile strength is shown in Fig. 9-7(*b*). It was observed that tensile strength increases as the compressive strength increases.

The influence of compressive strength on maximum crack width w_c is shown in Fig. 9-7(*c*). The trend of the variation of w_c via compressive strength is similar to that of fracture energy via strength.

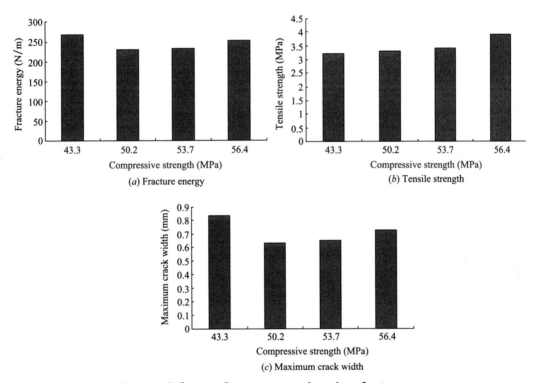

Fig. 9-7 **Influence of concrete strength on the softening curve**

9.5 Conclusions

Based on the stable fracture experiments of 12 groups of TPB specimens, the tetralinear softening curves of concrete were obtained by inverse analysis. The effects of maximum coarse aggregate size, specimen size and concrete strength on the softening curve were preliminary explored. Some conclusions were drawn as follows:

(1) The fracture energy, tensile strength and maximum cohesive crack width increased with the increase of the aggregate size. However, there was a downward trend of the tensile strength when the

aggregate size reaches a certain value. (2) The fracture energy as well as maximum crack width appeared the asymptotic behavior with the variation of specimen size. The specimen size which has little impact on the tensile strength of concrete was observed. (3) The tensile strength increased as the compressive strength increased. The trend of the variation of maximum crack width via compressive strength was similar to that of fracture energy via strength which showed no clear trend.

References

[1] Lin C, Jin X Y, Li Z J. Experimental investigation of effects on fracture characteristics of concrete [J]. China Concrete and Cement Products, 2004, (5):7-9

[2] Kleinschrodt H D, Winklen H. The influence of the maximum aggregate size and size of specimen on fracture mechanics prarmeters[C]. Fracture toughness and fracture energy of concrete, The Netherlands; Elsevier Science Publishers BV, 1986: 391-402.

[3] Xu S, Zhao G F. Fracture energy of concrete and its variation trend in size effect studied by using three-point bending beams [J]. Journal of Dalian University of Technology, 1991, 31(1):79-86

[4] Qian J S, Fan Y R and Yuan J... Size effect of the fracture energy of concrete tested by three point bending [J]. Journal of Chongqing Jianzhu University, 1995, 17(2): 1-8.

[5] Kwon S H, Zhao Z, Shah S P. Effect of specimen size on fracture energy and softening curve of concrete: Part II. Inverse analysis and softening curve [J]. Cement & Concrete Research, 2008, 38(8-9):1061-1069.

[6] Luan S, Darwin D. Effect of age and strength on fracture energy of concrete [J]. Journal of Hydraulic Engineering, 1999, (10):29-32

[7] Zhao Z, Yang L, Zhao Z, et al. Test Data Processing Method of Fracture Experiments of Dam Concrete for Inverse Analysis[M]. Computational Structural Engineering. Springer Netherlands, 2009, (22-24):1239-1247.

[8] Hillerborg A, Modéer M, Petersson P E. Analysis of crack formation and crack growth in concrete by means of fracture mechanics and finite elements [J]. Cement & Concrete Research, 1976, 6(6):773-781.

[9] Gopalaratnam V S, Ye B S. Numerical characterization of the nonlinear fracture process in concrete [J]. Engineering Fracture Mechanics, 1991, 40(6):991-1006.

[10] Uchida Y, Barr B I G. Tension softening curves of concrete determined from different test specimen geometries[C]. Fracture Mechanics of Concrete Structures. Freiburg: Aedificatio Publishers, 1998:387-398.

[11] Mihashi H, Nomura N, Niiseki S. Influence of aggregate size on the fracture process zone of concrete detected with three dimensional acoustic emission technique. Cement and Concrete Research, 1991, 21:734-744

[12] Wittmann F.H., Mihashi H. and Nomura N.. Size effect of fracture energy of concrete [J]. Engineering Fracture Mechanics, 1990, 35(1-3): 107-115

[13] Zhao Z, Kwon S H and Shah S P. Effect of specimen size on fracture energy and softening curve of concrete: Part I. Experiments and fracture energy [J]. Cement & Concrete Research, 2008, 38(8-9):1049-1060.

[14] Duan K, Hu X Z, Wittmann F H. Size effect on specific fracture energy of concrete [J]. Engineering Fracture Mechanics, 2007, 74(1):87-96.

10 Softening Behaviors of Dam Concrete and Wet-screening Concrete

10.1 Introduction

The total concrete placement on the mainframe of Three Gorges Project is more than 27 million m^3, and the research on the concrete fracture properties of its spillway section will provide a basis for the prevention and control of cracks during the construction period and the analysis of cracks during the operation period, which is of great significance for the later observation of cracks and the anti-seismic safety evaluation. Currently, there are two major approaches for the determination of concrete tension softening curve. One is the direct tension test[1-3], which is an ideal method used to determine the concrete softening curve, but it poses a higher requirement for the testing machine and needs to solve the loading alignment and other problems, with a higher degree of difficulty. The other is the inverse analysis[4-7]. P-CMOD curve is measured using wedge splitting test, through numerical simulation. The softening curve can be acquired using inverse analysis. The inverse analysis is a user-friendly, economical and time-efficient method. However, it also has some deficiencies: due to the existence of multiple local optimal solutions in numerical calculation, the objectivity of inverse analysis has been controversial. This research performed the wedge splitting test on 19 groups of dam concrete and wet-screening concrete specimens and the direct tension test on 8 groups of the corresponding material specimens. The softening curves of dam concrete and wet-screening concrete were achieved by inverse analysis calculation based on the improved evolutionary algorithms. This research introduced local search mechanism into the evolutionary algorithm[8], and addressed the controversial objectivity of inverse analysis by modifying the optimization method during the inverse analysis. In addition, it tested the validity of the adopted inverse analysis through an appropriate comparison of the results from direct tension test and those from inverse analysis. Meanwhile the softening behaviors of dam concrete and wet-screening concrete were analyzed hereby.

10.2 Experiments

10.2.1 Mix propotions

The mixture ratio of concrete materials used in this experiment is listed in Table 10-1 (the

10 Softening Behaviors of Dam Concrete and Wet-screening Concrete

Mix proportions of dam concrete and wet-screening concrete   Table 10-1

DT group	WS group	Concrete Grade	Gradation	Maximum Aggregate size(mm)	Mix parameters				Unit weight(kg/m³)							
					w/b	Unit consumption of water(kg/m³)	Fly ash ratio (%)	Sand Ratio (%)	Cement	Fly ash	Sand	Small gravel	Medieum gravel	Large gravel	Water reducer	Air entraining agent
DT2	WS12-12y	C20	One	10	0.50	140	30	45	196	84	869	1090	/	/	1.680	0.0196
DT4	WS13-17	C20	One	20	0.50	132	30	43	185	79	846	1152	/	/	1.584	0.0185
DT5	WS18	C30	One	20	0.45	135	20	42	240	60	814	1154	/	/	1.800	0.0195
DT6	WS19	C40	One	20	0.35	135	20	40	309	77	744	1145	/	/	2.316	0.0251
DT7	WS20	C50	One	20	0.30	140	10	39	420	47	698	1121	/	/	2.802	0.0280
DT3	WS21	C20	Two	40	0.50	120	30	38	168	72	769	515	772	/	1.440	0.0168
DT8a	WS22-25	C20	Three	80	0.50	102	30	31	143	61	653	373	373	745	1.224	0.0143
DT8	WS32-35	C20	Three	80	0.50	102	30	31	143	61	653	373	373	745	1.224	0.0143

Note: According to the Test Code for Hydraulic Concrete[9], wet-screening concrete is to remove all the aggregate which size is larger than 40mm from fresh dam concrete.

same as the materials used in the spillway section of Three Gorges Project). Of which, the cement was SANXIA 42.5 moderate heat portland cement produced by Gezhouba Cement Factory, the fly ash was Grade I fly ash produced by Shandong Zouxian Power Plant, the coarse aggregate was crushed gravel from Gushuling of Three Gorges, and the fine aggregate was crushed sand from Xiaanxi of Three Gorges, which were mixed using the drinking water from Three Gorges. The testing age of specimens is about one year.

10.2.2 Specimens

The geometry of direct tension specimen (DT) and wedge splitting specimen (WS) is schematically illustrated in Fig. 10-1. Among them the dimension of natural gradation dam concrete DT specimen is $B \times D \times L$=250mm×250mm×500mm, dimension of the other DT specimens is $B \times D \times L$=150mm×150mm×300mm. There are totally 8 DT specimen groups and 5 specimens in each group. However, each group would contain abnormal or untested specimens, and after the removal of such specimens, the rest for each group is listed in Table 10-2. There are 4 specimens in each WS group (see Table 10-3). The a_0 is initial crack length. Meanwhile the basic mechanical properties of the same batch concrete were tested.

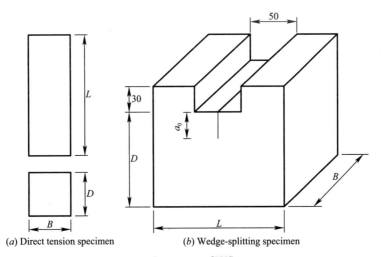

(a) Direct tension specimen (b) Wedge-splitting specimen

Fig. 10-1 Geometry of WS specimen

Direct tension specimens Table 10-2

Specimen group	Specimen				
DT[2]	DT6	/	DT8	DT9	DT10
DT[4]	/	DT17	/	/	DT20
DT[5]	DT21	DT22	DT23	DT24	DT25
DT[6]	DT26	/	DT28	DT29	DT30
DT[7]	/	DT32	DT33	DT34	DT35
DT[3]	DT11	DT12	DT13	/	/

10 Softening Behaviors of Dam Concrete and Wet-screening Concrete

continue

Specimen group			Specimen		
DT[8a]	DT56	DT57	/	/	DT60
DT[8]	/	DT37	DT38	DT39	DT40

WS specimens Table 10-3

Specimen	Aggregate category	Max. Aggregate size (mm)	Specimen size (mm)				Companion specimen	Effective specimen
			B	D	L	a_0		
WS12	Small aggregate	10		300	300	150	4	3
WS12y				600	600	300	4	3
WS13	Small aggregate primary gradation	20	200	300	300	150	4	3
WS14				600	600	300	4	3
WS15				800	800	400	4	3
WS16				1000	1000	500	4	4
WS17				1200	1200	600	4	3
WS18				300	300	150	4	4
WS19				300	300	150	4	2
WS20				300	300	150	4	3
WS21	Two gradation	40		300	300	150	4	3
WS22	Fully-graded aggregate dam concrete	80	250	450	450	225	4	3
WS23				800	800	400	4	3
WS24				1000	1000	500	4	3
WS25				1200	1200	600	4	3
WS32	Wet-screening concrete	40	200	300	300	150	4	3
WS33				600	600	300	4	3
WS34				800	800	400	4	3
WS35				1000	1000	500	4	3

10.2.3 Direct tension method

Direct tension test was conducted in National Laboratory of Large Dam Structures, Tsinghua University. This experiment used Instron 8506 300t digital triaxial electro-hydraulic servo testing machine with the rigidity of 6000kN/mm, which is a large-rigidity electro-hydraulic servo closed-circuit testing system tailored for the testing of high-strength materials. This testing system ball hinge has two ball hinges in the upper and lower part respectively, with a loading capacity up to 3000kN. The deflection angle of each ball hinge is 20 degrees, so that the relative eccentricity is kept less than 10%. It used a microcomputer to provide control signals for the closed circuit system and collect data. In 48 hours before the experiment, the upper and lower plate was bonded using epoxy resin. When the structural adhesive reaches certain strength, in order to assure the axial alignment of loads, the specimens were connected with the upper, lower ball hinge (see Fig. 10-2), and the specimens should be carefully placed with the loading alignment to avoid creating larger eccentric loads. To measure the axial deformation of specimens, 4 displacement meters were arranged within the entire altitude and the measuring range of each displacement meter was ±5mm. The average value of the readings from these four displacement meters was used to calculate the strain of specimens. To measure the stress-strain curve for each

Fig. 10-2 Direct tension test

specimen, the maximum value among the readings of these 4 displacement meters was used as feedback signals for loading control. Throughout the experiment, the strain control loaded was at 5με/min to ensure the stable operation of direct tension test. The loading and data acquisition were controlled using a computer, where the acquisition frequency was 0.2pt/s, and the initiation and propagation of cracks were observed while loading.

10.2.4 Wedge splitting test method

The wedge splitting test was performed in a very stiff testing machine. The support for this experiment was provided at one fourth of the specimen width L and the wedge angle was set as 15°. The loading device and force applied are as shown in Fig. 10-3: P_v is the applied vertical force and P_h is the horizontal force applying to specimen by the force transfer device. A 100 kN capacity load cell was used to measure the applied load, and the accuracy was ±2% of the maximum applied load. The crack mouth opening displacement (*CMOD*) was measured with a CDP-5 displacement sensor made in Japan, which has a capacity and accuracy of 5mm and 0.0005mm. The test was controlled by a constant *CMOD* rate of 0.15mm/min. The fracture experiment was steadily performed, and two 8-channel DH5937 dynamic strain testing systems worked jointly to acquire the test data of load and deformation during the whole process. The *CMOD* value was recorded until the specimen was broken into two halves. The *P-CMOD* curves of WS specimens were achieved by the tests.

10 Softening Behaviors of Dam Concrete and Wet-screening Concrete

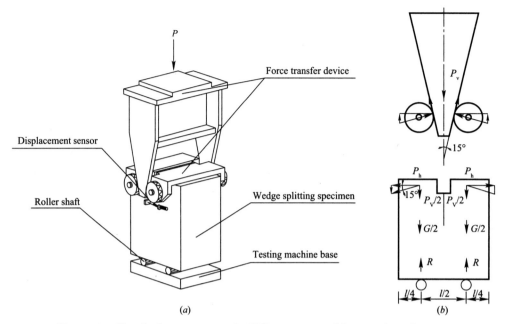

Fig. 10-3 Sketch of test apparatus for WS specimen and force analysis diagram

10.3 Softening Curves of Dam Concrete and Wet-screening Concrete Determined by the Direct Tension Test Method

The determination of concrete softening curve through direct tension test was explained by illustrating the direct tension test DT[3]. The test results of dam concrete and wet-screening concrete can be obtained by the DT tests as follows:

(1) Original load-strain curve (P-ε curve), as shown in Fig. 10-4.

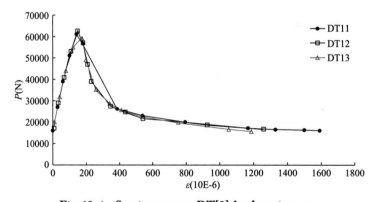

Fig. 10-4 Specimen group DT[3] load-strain curve

(2) Stress-strain curve (σ-ε curve), as shown in Fig. 10-5.

10.3 Softening Curves of Dam Concrete and Wet-screening Concrete Determined by the Direct Tension Test Method

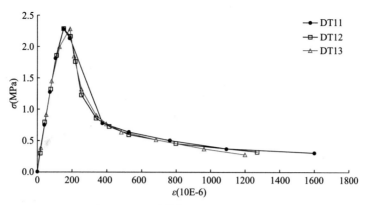

Fig. 10-5 DT[3] stress-strain curve

(3) Stress-deformation curve (σ-δ curve), as shown in Fig. 10-6.

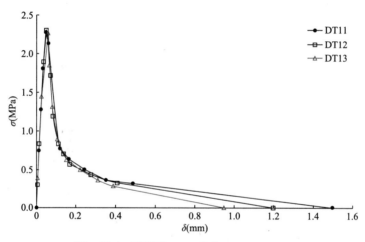

Fig. 10-6 DT[3] stress-deformation curve

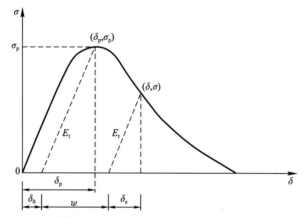

Fig. 10-7 Concrete stress-strain curve

(4) Stress-crack opening width curve (σ-w curve). In the direct tension test, the total deformation of concrete δ comprises the following parts:

$$\delta = \delta_e + \delta_0 + w \qquad (10\text{-}1)$$

The elastic deformation δ_e can be calculated according to the following formula:

$$\delta_e = \sigma l / E_t \qquad (10\text{-}2)$$

Where σ_p is the peak stress, σ_{ep} is the elastic deformation. When the stress reaches to the peak value, E_t is the tensile elastic modulus, and l is the specimen length.

The residual deformation δ_0 can be calculated according to the following formula:

$$\delta_0 = \delta_p - \delta_{ep} \qquad (10\text{-}3)$$

Where δ_p is the specimen deformation when the stress reaches to the peak value.

The crack opening width w can be obtained based on stress-deformation curve (σ-δ curve) of DT specimen.

$$w = \delta - \delta_e - \delta_0 \qquad (10\text{-}4)$$

The σ-w curve of DT[3] was achieved as shown in Fig. 10-8.

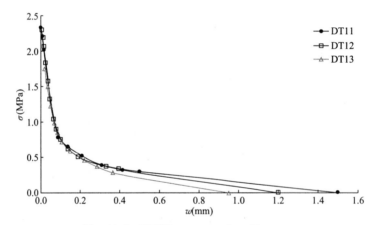

Fig. 10-8 DT[3] stress-crack width curve

(5) The relative stress-crack width curves (σ/f_t-w/w_0 curve) is shown in Fig. 10-9.

(6) Tensile strength f_t, maximum crack width w_0, fracture energy G_F and analytic expression of the softening constitutive relation σ-w curve.

By fitting the data of σ-w curve of three specimens in the group DT[3] as shown in Fig. 10-9, the parameters f_t, w_0, G_F were obtained as 2.341MPa, 1.22mm, and 401.394N/mm respectively. The expression of softening relation of DT[3] was achieved by the least square method as follows:

$$\sigma = f_t \left\{ 1 - \varphi \exp\left[-\left(\frac{\lambda}{\frac{w}{w_0}} \right)^n \right] \right\} \qquad (10\text{-}5)$$

Where φ, λ, n are material parameters. For the specimen group DT[3], $\varphi=1.2$, $\lambda=0.03$, $n=0.48$, $R^2=0.9845$.

10.4 Experimental Results of Wedge Splitting Test and Test Data Processing for Inverse Analysis

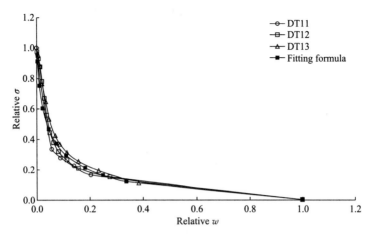

Fig. 10-9 Relative DT stress-crack width curves of specimen group DT[3]

Similarly, the softening relation expression and softening curve control parameters for other specimen groups can be obtained, and the final results are given in Table 10-4.

Parameters of the softening curve expression of DT specimens Table 10-4

Category	Specimen group	Maximum aggregate size d_{max} (mm)	Concrete strength-grade	φ	λ	n	R^2	f_t (MPa)	w_0 (mm)	G_F (N/mm)
Small aggregate primary gradation	DT[2]	10	C20	1.200	0.030	0.480	0.9743	2.483	0.75	260.309
	DT[4]	20	C20	1.230	0.030	0.470	0.9743	2.455	0.90	278.213
	DT[5]	20	C30	1.230	0.045	0.530	0.9799	2.103	0.89	303.236
	DT[6]	20	C40	1.150	0.040	0.620	0.9800	2.425	0.79	262.002
	DT[7]	20	C50	1.210	0.035	0.480	0.9911	2.298	0.80	272.218
	DT[3]	40	C20	1.200	0.030	0.480	0.9845	2.341	1.22	401.394
Large aggregate Natural gradation	DT[8a]	80	C20	1.210	0.070	0.600	0.9831	1.394	1.55	370.298
Wet-screening Natural gradation	DT[8]	40	C20	1.150	0.040	0.580	0.9754	2.314	0.92	341.298

10.4 Experimental Results of Wedge Splitting Test and Test Data Processing for Inverse Analysis

10.4.1 Preliminary processing

Because the wedge splitting test recorded a very large amount of test data for each specimen, for example, the measured data of *P-CMOD* curve was up to 30000-300000 lines, they should be initially processed before solving the softening curve in the inverse analysis, and the *P-CMOD* curve for reflecting the fracture performance of each specimen should be obtained in an unbiased way.

The measured data was initially processed by taking the following three steps: (1) Remove all data points with negative horizontal splitting load; (2) Remove the data points far away from *P-CMOD* trend line; (3) Remove *P-CMOD* measured results accompanying with the specimens that were greatly different from other specimens in the same group or with very unsmooth curve.

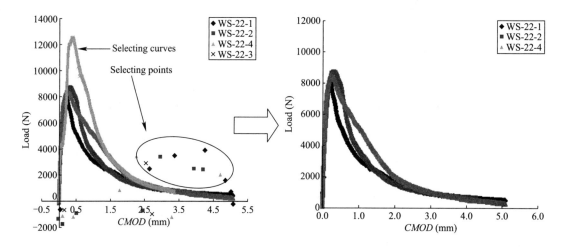

Fig. 10-10 Preliminary data processing diagram

Through the above initial processing, *P-CMOD* curves for some specimens are as shown in Fig. 10-11. *P* denotes the horizontal splitting force as shown in Fig. 10-10, $P=0.5P_v/\tan\alpha$, where $\alpha=15°$, then $P\approx1.866P_v$, where P_v denotes the vertical load applied by the testing machine, as shown in Fig. 10-3.

10.4.2 Data smoothing

One reference[10] suggests that the *P-CMOD* curve testing results accompanying each specimen group should be processed into a representative curve, after the initial processing of measured data. However, Fig. 10-11 indicates a difference between experimental results accompanying each specimen in the same group. This would affect the inverse analysis result to some extent. Thus, This research was intended to make an independent inverse analysis calculation for each specimen. After that, it performed the average fitting for the softening curve obtained from inverse analysis of a specimen in each group, which was used as the softening curve for this specific specimen.

Because the experiment data after initial processing was very large, the inverse analysis was not applicable to it, which would affect the calculation efficiency of inverse analysis; and it can be seen from Fig. 10-11 that the *P-CMOD* curve for individual specimen was not smooth, such experimental data would make it difficult for the inverse analysis calculation, and therefore, the measured data of wedge splitting specimen needs to be further processed.

10.4 Experimental Results of Wedge Splitting Test and Test Data Processing for Inverse Analysis

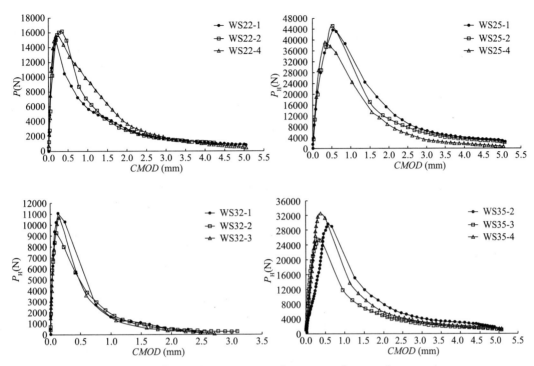

Fig. 10-11 Some P-$CMOD$ curve of specimen after initial treatment

After initial processing, the P-$CMOD$ curves were set in smoothened using quadratic spline approximation[11] to obtain the representative curve of around 200 points, which will be used as the input data of the inverse analysis. In addition, since the shape of the measured P-$CMOD$ curves may vary from specimen to specimen, a smoothing parameter λ was introduced to manually adjust the fitting results, so as to obtain the appropriate P-$CMOD$ representative curve.

The further processing of measured data of wedge splitting specimen was explained by illustrating the WS25-1 specimen as shown in Fig. 10-11. The smoothing parameter λ is a key parameter during the processing, with a value ranging from 1E-2 to 1E+8, the best representative curve for P-$CMOD$ curves after initial processing was obtained by manually adjusting λ, which was prepared for the inverse analysis calculation.

As shown in Fig. 10-12, the measured data for WS25-1 was not smooth. If λ=1E+8, although two P-$CMOD$ curves are fit well, the results yielded from the originally measured P-$CMOD$ curves are not very good. It would definitely affect the result and efficiency of inverse analysis, provided that the fitting P-$CMOD$ curve when λ=1E+8 was used as the input data for inverse analysis. In this case, the value of λ may be manually modified; the results when λ=1E+2 are as shown in Fig. 10-13; it is still acceptable despite the greater difference between them. Therefore, the P-$CMOD$ representative curve as shown in Fig. 10-13 was used as the input data of inverse analysis calculation for WS25-1.

It is worth noting that more attention should be paid in selecting the value of λ. The smoother curve is not necessarily good, and it should be subject to the measured P-$CMOD$ curve with the least error to the extent possible.

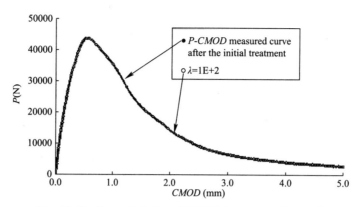

Fig. 10-12 Smooth fitting of WS25-1 specimen($\lambda=1E+8$)

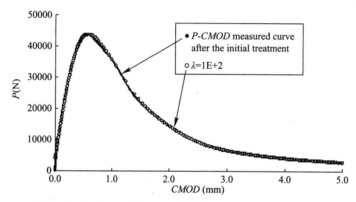

Fig. 10-13 Smooth fitting of WS25-1 specimen($\lambda=1E+2$)

10.5 The Softening Curve of Dam and Wet-screening Concrete Determined by the Evolutionary Algorithm-based Inverse Analysis Method

This research compiled the source program to solve the softening curves of dam and wet-screening concrete through the evolutionary algorithm-based inverse analysis. The flowchart is as shown in Fig. 10-14. The initial input data for the inverse analysis calculation include: the geometric dimensions and weight of specimen, the compressive strength of material f_c and the tangent modulus elasticity of origin E, and the initial setting of softening parameter. Moreover, the meshing form, the error function and the optimization algorithm used during the inverse analysis would greatly impact the accuracy and efficiency of the inverse analysis calculation. In addition to the wedge splitting fracture test, our research team also performed the supporting concrete direct tension test and the associated test of basic mechanical property. The fracture parameters measured in these two tests were referenced in the parameter setting of the inverse analysis.

10.5.1 Original input data

It is indicated that the weight of specimen used in the wedge splitting test has a very small impact on the inverse analysis[12], and therefore, the weight of specimen was set as 0 in the inverse analysis calculation.

The tangent modulus E is the basic mechanical parameter of concrete material, which has a greater impact on the inverse analysis calculation. The secant elastic modulus E_c can be measured through the basic mechanical property test and the required tangent modulus E would be obtained through conversion.

In accordance with the hydraulic concrete testing procedures, the prism compressive strength f_c and the secant modulus E_c of the concrete material were tested and their results are listed in Table 10-5.

The formula of secant elastic modulus E_c can be expressed as follows:

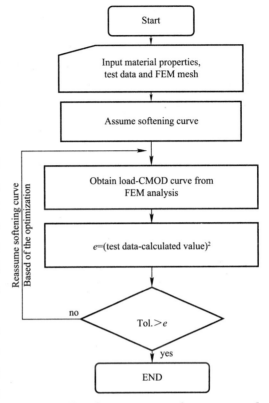

Fig. 10-14 Based on optimization dam concrete and wet-screening concrete nverse analysis of softening curve chart

$$E_c = \frac{P_2 - 1}{A} \times \frac{L}{\Delta L} \quad (10\text{-}6)$$

Where P_2 is 40% of the ulimate load, P_1 is the load when the stress reaches to 0.5MPa, L is the gauge length, ΔL is deformation increment from load P_1 to P_2. and A is the cross sectional area.

As shown in Fig. 10-15, the secant elastic modulus E_c denotes the secant modulus from P_1 (approximate at the origin) to P_2. The elastic modulus required for the inverse analysis should be the tangent modulus E.

In accordance with the European Concrete Model Specification[13] (CEB-FIP-90), the tangent modulus E can be deduced using the following equation (10-7).

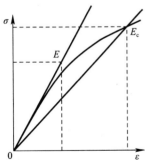

Fig. 10-15 Schematic diagram of the correlation between tangent modulus of original point and secant modulus

$$E = E_c / 0.85 \quad (10\text{-}7)$$

Therefore, the secant modulus E_c cannot be directly used to perform the inverse analysis, or it may lead to the distortion of inverse analysis result.

The elastic modulus obtained by equation (10-7) are listed in Table 10-5.

Table 10-5 Material characteristics of wedge splitting specimens and the corresponding DT test results

Specimen	f_c (MPa)	E_c (GPa)	E (GPa)	f_t (MPa)	w_c (mm)	G_F (N/m)
WS12	30.3	26.7	31.4	2.483	0.75	260.3
WS12y	30.3	26.7	31.4	2.483	0.75	260.3
WS13	34.2	33.4	39.3	1.886	0.76	261.4
WS14	34.2	33.4	39.3	1.886	0.76	261.4
WS15	34.2	33.4	39.3	1.886	0.76	261.4
WS16	33.0	33.4	39.3	1.886	0.76	261.4
WS17	34.2	33.4	39.3	1.886	0.76	261.4
WS18	38.6	29.6	34.8	2.103	0.89	303.2
WS19	43.3	30.5	35.9	2.425	0.79	262.0
WS20	55.4	34.8	40.9	2.298	0.80	272.2
WS21	31.5	31.5	37.1	2.341	1.22	401.4
WS22	28.5	28.0	32.9	1.394	1.55	370.3
WS23	34.4	29.1	34.2	1.394	1.55	370.3
WS24	34.4	29.1	34.2	1.394	1.55	370.3
WS25	28.5	28.0	32.9	1.394	1.55	370.3
WS32	34.4	29.1	34.2	2.314	0.92	341.3
WS33	34.4	29.1	34.2	2.314	0.92	341.3
WS34	28.5	28.0	32.9	2.314	0.92	341.3
WS35	28.5	28.0	32.9	2.314	0.92	341.3

As the research used the evolutionary algorithm and the initial setting of softening parameters had no significant impact on the efficiency and result of inverse analysis calculation, only a rough initial value was needed to input.

10.5.2 Principle of FEM simulation of concrete crack propagation

During the crack propagation of wedge splitting specimen, it is assumed that those areas other than the fracture process zone (FPZ) were in elasticity. As shown in Fig. 10-16, the hinge constraint was set along the crack propagation direction. When the tensile stress of concrete along the crack propagation direction reached the tensile strength f_t, the hinge constraint of node was released. FEM simulation of crack propagation was accomplished by top-down release of hinge constraints one by one along the propagation direction. Based on the fictitious crack model (FCM), the non-linear fracture of concrete was converted into the equivalent linear elastic fracture[14] to simulate the crack propagation by applying the cohesive force on FPZ.

10.5.3 Mesh generation

The simulation was simplified to the plane stress calculation. The half of the specimen was analyzed in accordance with the symmetry. The mesh generation as shown in Fig. 10-17, if adopted, can significantly improve the calculation efficiency.

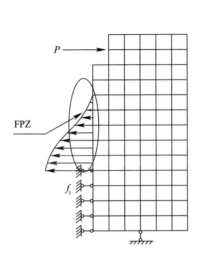

Fig. 10-16 Diagram of the simulation of concrete crack propagation based on fictitious crack model

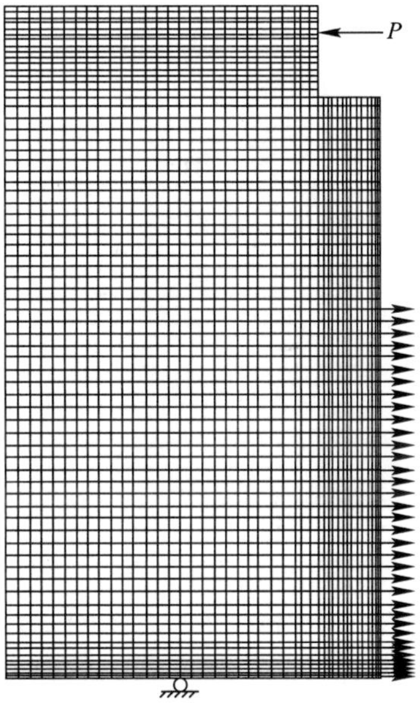

Fig. 10-17 Mesh generation

The fineness level of finite element meshing, especially the meshing at the fracture ligament, would have a greater impact on the calculation efficiency and the objectivity of inverse analysis result. Our research team did a detailed research on this concern, Yang Lijian[10] proposed the meshing form that encrypts on both ends of fracture ligation in his work and verified the applicability of this experiment. Therefore, this research used a similar meshing form and provided three different meshing densities for the convenience of research, namely fine, normal and coarse.

The simulation of crack propagation is based on the FCM model. Namely, from the start of precast crack tip, when the stress of crack tip reaches the tensile strength, the node of crack tip will gradually move along the crack propagation path, and the corresponding external load and displacement value can be also obtained.

10.5.4 Error function e

The selection of error function is the precondition for any optimization problem, and the error function also has a greater impact on the final optimization result. Given that an inverse analysis needs to calculate the error, the discrete measured P-CMOD curve needs to be firstly approximated to be a continuously smooth curve. Since the discrete degree of measured curve data may vary from group to group, the two-stage spline approximation of adjustable smoothing parameters was utilized to complete this work. The error function has three forms:

(1) Average relative error e_1

$$e_1 = \sum_{i=1}^{n} \frac{|\arctan(m_i / k)|}{90°} \cdot \left|\frac{f_i - y_i}{m_i \bar{x}}\right| + \left\{1 - \frac{|\arctan(m_i / k)|}{90°}\right\} \cdot \left|\frac{f_i - y_i}{\bar{y}}\right| \quad (10\text{-}8)$$

Where (x_i, y_i) are the points of calculated P-CMOD curve, and (x_i, f_i) are the points of measured P-CMOD curve.

Where $y=Ax^2+Bx+C$, and A, B, C were determined by the points $P_{i-1}(x_{i-1}, y_{i-1})$, $P_i(x_i, y_i)$ and $P_{i+1}(x_{i+1}, y_{i+1})$.

$$m_i = \left.\frac{dy}{dx}\right|_i, \forall i \in [2, n-1]; m_1 = \frac{y_2 - y_1}{x_2 - x_1}, i=1; m_n = \frac{y_n - y_{n-1}}{x_n - x_{n-1}}, i=n$$

$\dfrac{|\arctan(m_i / k)|}{90°}$ is for a weighted error measure. If the k is small ($k \approx 1$), e_1 gives the relative x-deviation. If the k is very large which is compared to the Young's modulus E, it passes into the relative y-deviation. If $k \approx E$, the x-deviation and y-deviation are almost equally weighted. Theoretically, the more brittleness of material means the larger value of k should be selected to achieve the better convergence performance.

(2) Maximum relative error e_2

$$e_2 = \max\left\{\frac{|\arctan(m_i / k)|}{90°} \cdot \left|\frac{f_i - y_i}{m_i \cdot \bar{x}}\right| + \left\{1 - \frac{|\arctan(m_i / k)|}{90°}\right\} \cdot \left|\frac{f_i - y_i}{\bar{y}}\right|\right\} \quad (10\text{-}9)$$

(3) Error of the fracture energy e_3

$$e_3 = \frac{1}{2}\left(\frac{G_{F\exp} - G_{F\text{num}}}{G_{F\exp}} + \frac{G_{F\exp} - G_{F\text{num}}}{G_{F\text{num}}}\right) \quad (10\text{-}10)$$

$G_{F\exp}$, $G_{F\text{num}}$ is the fracture energy obtained from the measured data and numerical simulation respectively, such equation took an account of the possible emergence of $G_{F\text{num}} \to 0$, $G_{F\text{num}} \to \infty$.

The e_1+e_3 error measure is generally recommended, but if the measured P-CMOD curve is incomplete, e_3 will not be used as an error measure.

10.5.5 Optimization algorithm

Over these years, scholars have studied the optimization of softening parameters. Roelfstra and Wittmann[4] firstly proposed an algorithm to reduce the deviation between the simulated P-CMOD curve and the measured one, and developed the corresponding software SOFTFIT.

10.5 The Softening Curve of Dam and Wet-screening Concrete Determined by the Evolutionary Algorithm-based Inverse Analysis Method

This software adopted the bilinear softening curve, which is based on the finite element method and the discrete crack model. However, its efficiency in determining the softening curve has not significantly improved if compared with manual fitting. Furthermore, such algorithm needs to make "the first guess" for the input softening curve, which means that it needs the result that is very close to the loop fitting as the input data, and only in this way could the risk arising from the local minimum of error function be avoided.

Japan Concrete Institute[15] proposed an algorithm that determines the softening curve of concrete based on a totally different concept. It firstly assumes a multi-segment linear softening curve, and the individual slopes are determined successively by adjusting a corresponding increment of the calculated *P-CMOD* curve to the tested one. In this way, the softening curve is gradually formed when simulating the crack propagation. The multi-segment linearity of softening curve can approximately exploit the high performance of the optimization method without an initial assumption about the shape of softening curve. However, the multi-segment linear approximation needs to assume the initial cohesive stress as the starting point of this softening curve, which diminishes the objectivity of inverse analysis result.

The above research indicates that the primary consideration for the selection of optimization algorithm is to avoid the local minimum, obtain the result that suits the physical significance, and improve the calculation efficiency. Therefore, the author argued that the stochastic optimization method is the most suitable option for the research purpose. The simulated annealing algorithm is a common stochastic optimization algorithm, but it has a slow convergence and may easily get into the local minimum. To overcome these disadvantages, the evolutionary algorithm was employed as the optimization method for inverse analysis.

As one of the artificial intelligent algorithms, evolutionary algorithm evolves from the biological evolutionary theory, which has been an extensively applied and efficient random searching and optimization method. It is an dynamic integration of "genetic algorithm" and "evolutionary strategy". It has these main advantages: (1) Evolutionary algorithm starts searching from the string set of solutions rather than from a single solution, which is significantly different from the conventional optimization algorithm. Therefore, evolutionary algorithm has a large coverage that benefits the global optimization. (2) Evolutionary algorithm evaluates multiple solutions in the searching space, thus reducing the risk of catching the local minimum. And it can improve the convergence rate and reduce the risk of obtaining the local minimum., These advantages are urgently needed for solving the optimization problems.

In this research, the local search mechanism was introduced to evolutionary algorithm, referred to as the cultural memetic algorithm.[16] That is to say, a special local search method was introduced in consideration of the similarity of parameter vector, which is also referred to as neighborhood attraction[17] according to a neural network learning strategy. The improved evolutionary algorithm can further accelerate the convergence rate and reduce the risk of getting into the local minimum.[18, 19]

To ensure the calculation efficiency and rationality of inverse analysis solution, this research

10 Softening Behaviors of Dam Concrete and Wet-screening Concrete

used the improved evolutionary algorithm to conducting the inverse analysis to achieve the softening curves of dam concrete.

10.5.6 Softening curves of dam concrete calculated by the inverse analysis method

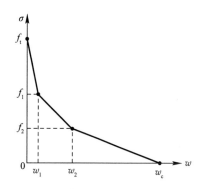

Fig. 10-18 Tri-linear softening curve

Average of softening curves

Based on the above-mentioned procedures, the trilinear softening curve of every specimen in each specimen group was obtained by inverse analysis method. The schematic diagram of a trilinear softening curve is shown in the Fig. 10-18.

Table 10-6 lists the control parameters of the tri-linear softening curves of specimen group WS12 obtained by inverse analysis method.

Softening curves of specimen group WS12 by inverse analysis method   Table 10-6

Specimen	f_t (MPa)	f_1 (MPa)	f_2 (MPa)	w_1 (mm)	w_2 (mm)	w_c (mm)
WS12-1	2.546	2.052	0.472	0.0083	0.1130	0.4907
WS12-3	2.203	1.601	0.412	0.0025	0.1173	0.4368
WS12-4	2.531	1.704	0.434	0.0020	0.1225	0.4688

This research averaged several softening curves of each specimen group to obtain the average softening curve representing the fracture properties of this specimen group. Take two softening curves as an example, an average fitting method of softening curves is illustrated in Fig. 10-19.

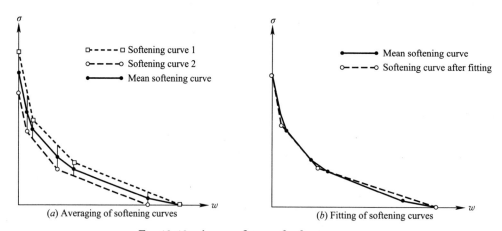

(a) Averaging of softening curves   (b) Fitting of softening curves

Fig. 10-19   Average fitting of softening curves

As shown in Fig. 10-19(a), the average method of softening curves was used to average the stress value σ corresponding to each inflection point of softening curves in the same specimen

10.5 The Softening Curve of Dam and Wet-screening Concrete Determined by the Evolutionary Algorithm-based Inverse Analysis Method

group, while the value of w_c was subject to the maximum value of softening curve in each specimen group. The fitting method for the averaged softening curve was achieved through the software named 1st Opt, as shown in Fig. 10-19(b). In essence, it is a process of optimization and the main idea is to assure the area under the two multi-segment softening curves being consistent to with each other.

Taking the above-mentioned specimen group WS12 as an example, the average fitting process of its softening curve is shown in the Fig. 10-20. It can be seen that the fracture energy G_F of the average softening curve and the fitting trilinear softening curve was 211.7N/m and 216.2N/m respectively, which are the satisfactory results.

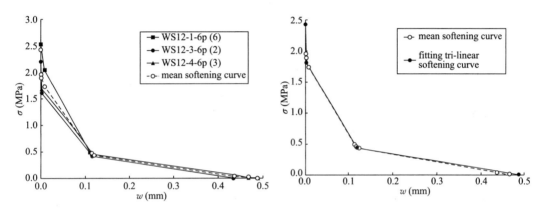

Fig. 10-20 Averaging and fitting of softening curve for specimen group WS12

The comparisons between the peak load (P_{max}) calculated based on the optimized softening curve and the tested one for every specimen, and those between the calculated CMOD at peak ($CMOD_c$) and the tested one are shown in the Fig. 10-21.

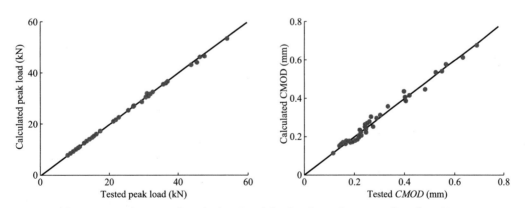

Fig. 10-21 Comparisons between the calculated peak load and tested one as well as those between $CMOD_c$

The Fig. 10-21 indicates that the calculated peak load coincides well with its measured value. Although the discreteness of the $CMOD_c$ is larger than that of P_{max}, they also coincide well with each other. This suggests that the optimization method based on the evolutionary algorithm is

123

feasible, which can better obtain the softening curves of dam and wet-screening concrete by the inverse analysis.

The parameters of softening curve and the fracture energy G_F of the dam concrete and wet-screening concrete by inverse analysis are listed in Table 10-7.

Parameters of trilinear softening curves and the G_F of wedge-splitting specimens   Table 10-7

Specimen	Aggregate size (mm)	Concrete strength grade	f_t (MPa)	f_1 (MPa)	f_2 (MPa)	w_1 (mm)	w_2 (mm)	w_c (mm)	G_F (N/m)
WS12	10	C20	2.427	1.808	0.433	0.0028	0.1184	0.4907	216.2
WS12y			2.337	1.576	0.400	0.0175	0.1619	0.7796	300.5
WS13			3.345	2.448	0.374	0.0118	0.1309	0.5688	283.9
WS14			4.109	1.726	0.155	0.0309	0.2808	1.4493	415.9
WS15		C20	3.812	1.711	0.201	0.0345	0.3581	1.4155	510.9
WS16	20		3.618	1.600	0.157	0.0324	0.3511	1.3111	439.7
WS17			4.313	2.137	0.075	0.0252	0.4449	1.5350	586.4
WS18		C30	3.816	1.724	0.253	0.0417	0.1688	0.7622	316.3
WS19		C40	2.733	2.032	0.159	0.0042	0.1506	0.6700	211.9
WS20		C50	3.724	1.785	0.117	0.0466	0.1968	0.6704	298.9
WS21	40	C20	3.261	1.517	0.186	0.0483	0.3246	1.0132	414.5
WS22			3.338	1.141	0.482	0.0255	0.2412	1.3186	551.9
WS23	80	C20	3.383	1.943	0.535	0.0171	0.2691	1.2899	630.8
WS24			3.349	2.070	0.277	0.0067	0.4433	1.5289	680.5
WS25			3.074	2.062	0.220	0.0051	0.4366	0.9269	559.6
WS32			2.990	1.541	0.235	0.0214	0.1813	0.6722	247.9
WS33	40	C20	3.566	1.529	0.294	0.0350	0.2576	0.9834	399.0
WS34			4.022	1.317	0.249	0.0380	0.3252	1.3363	452.5
WS35			3.191	1.440	0.246	0.0260	0.2986	1.2358	405.7

10.6 Comparison between the Softening Curves by Inverse Analysis Method and Those by Direct Tension Method

Our research group also performed the direct tension test of dam and wet-screening concrete associated with the wedge splitting specimen. A comparative study was made on the softening curves (three control parameters of each softening curve) of dam and wet-screening concrete obtained by inverse analysis and direct tension test, and some comparisons were selected as shown in Figs. 10-22, 10-23 and 10-24, respectively.

10.6 Comparison between the Softening Curves by Inverse Analysis Method and Those by Direct Tension Method

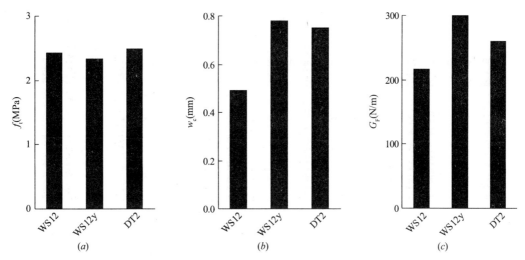

Fig. 10-22 Comparison of the control parameters of softening curve between the two methods for the small aggregate size specimens (d_{max}=10mm)

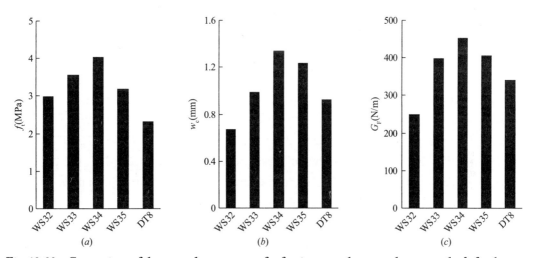

Fig. 10-23 Comparison of the control parameters of softening curve between the two methods for the wet-screening concrete specimens (d_{max}=40mm)

It can be seen from the above figure that when the aggregate size of specimen is 40mm or below, the softening curves obtained by inverse analysis and direct tension test are relatively consistent, which is also indicated from Fig. 10-22 and Fig. 10-23. As it is also the same case with other specimen groups, no description is given here. For the fully-graded aggregate dam concrete, the softening parameters obtained by these two methods are different. Since there is only one large aggregate specimen group of direct tension test, the further study about the cause to the above phenomenon needs to be made. This suggests that the results obtained by inverse analysis and direct tension test are substantially consistent, which verifies the feasibility of inverse analysis.

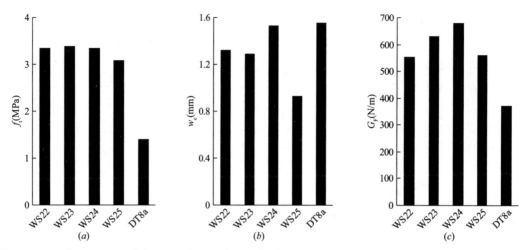

Fig. 10-24 Comparison of the control parameters of softening curve between the two methods for the large aggregate dam concrete specimens (d_{max}=80mm)

10.7 Softening Behaviors of Dam Concrete and Wet-screening Concrete

10.7.1 Effect of compressive strength on the softening curve

The tensile strength f_t, maximum cracking opening width w_c and fracture energy G_F are the three controlling parameters of concrete softening curve. They can characterize the crack resistance. The fracture energy G_F reflects the crack resistance from the perspective of energy field. This section compared three control parameters: the tensile strength f_t, the maximum cracking opening width w_c and the fracture energy G_F, of the softening curve obtained through inverse analysis from the wedge splitting specimens in d_{max}=20mm, the aggregate size: $B \times D \times L$=200mm×300mm×300mm, with the concrete compressive strength: 34.2MPa(WS13), 38.6MPa(WS18), 43.3 MPa(WS19)and 55.4MPa(WS20)respectively, as shown in Fig. 10-25.

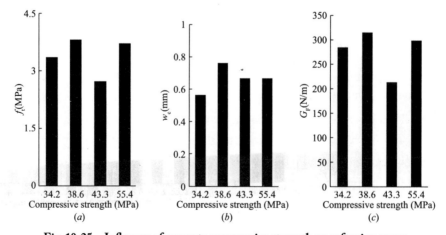

Fig. 10-25 Influence of concrete compressive strength on softening curve

It can be seen from the above figure that the trends of the compressive strength f_t and the fracture energy G_F is very similar, which increases and tends to be stable, with the increase of concrete compressive strength, and such trend is more obvious for the maximum cracking opening width w_c. To some extent, the strength and fracture energy of concrete characterizes the crack resistance, and therefore in theory, f_t and G_F should increase with the concrete compressive strength. However, the actual results were not the case, for this point, other scholars obtained the same finding in their studies, which can be described as: with the increase of concrete strength, there is a single case that the fracture energy may decrease.[20, 21] The possible reason may be primary attributed to: with the increase of concrete strength, the bonding strength between cement paste and aggregate will increase as well, but when the aggregate size is smaller, the bonding strength is often higher than the aggregate strength, and in this case, the concrete compressive strength depends on the aggregate strength. In this experiment, the aggregate size d_{max}=20mm, which is relatively smaller, with the increase of concrete strength, the concrete failure was firstly controlled by the aggregate strength, despite the continuous increase of concrete strength, the energy required for specimen failure didn't increase any more, but tended to be stable, with a single case of decrease.

10.7.2 Effect of aggregate size on the softening curve

The fracture properties of concrete depend on the mechanical performance of aggregate, hardened cement matrix and the bonding face between them. The effects of aggregate on fracture properties of concrete mainly include: (1) There is micro-cracks between aggregate and cement matrix, which may further develop into cracks, thus reducing the aggregate interlocking effect; (2) If the cracks propagate across the aggregate, then aggregate may hinder the crack propagation. This research compared three control parameters: the tensile strength f_t, the maximum crack opening width w_c and the fracture energy G_F, of the softening curves obtained through inverse analysis from the wedge splitting specimens (WS12, WS13 and WS20) with the same concrete strength ($f_c \approx$ 32MPa), in the same dimensions: $B \times D \times L$=200mm×300mm×300mm, with different aggregate sizes (d_{max}=10mm, 20mm and 40mm, respectively), as shown in Fig. 10-26.

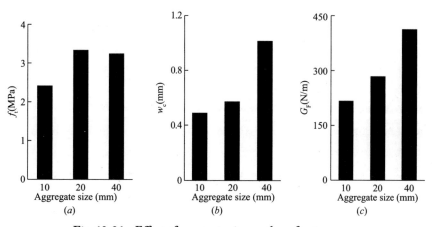

Fig. 10-26 Effect of aggregate size on the softening curve

Fig. 10-26 shows that the maximum cracking opening width w_c and the fracture energy G_F tends to increase with the aggregate size, but the tensile strength f_t increases then slightly decreases.

As to the change law of f_t, these findings are very similar to the findings by Hou Xinyue.[22] The tensile strength f_t is the largest when d_{max}=20mm, which indicates that the concrete tensile strength has the maximum size effect. Although the further study is needed, a reasonable explanation can be given as: when d_{max} is smaller, and if the cement content and sand ratio are fixed, then the content of cement paste in concrete also stays the same. When the slump is consistent, the concrete with smaller d_{max} will have larger water-cement ratio due to its large surface area, so its strength will decline; but when d_{max}=40mm: (1) With the increase of d_{max}, the bonding force between aggregate and cement matrix is diminished, thus reducing the concrete strength; (2) The shrinkage of cement matrix will result in the tensile stress in concrete, which is proportional to the aggregate size, so the concrete strength will decline.

It can be seen from Fig.10-26(c) that the fracture energy G_F will increase with the aggregate size, which is consistent with the findings by Zhang et al.[23] In general, the bonding force between aggregate and cement matrix in concrete is relatively weak. Depending on the aggregate failure or not, the cracks on concrete have two ways of propagation: (1) If the aggregate cracking, the cracks propagate into the aggregate; (2) If the aggregate does not crack, the cracks propagate along the bonding face, and there is an obtuse angle between the crack propagation direction and the original crack.[24] As it is known to all, since the aggregate will hinder these two ways of propagation, the larger aggregate size means more winding cracks and larger fracture ligament area, so that the fracture energy G_F increases.

10.7.3 Effect of specimen size on the softening curve

This research examined the size effect of softening curve from view of the ligament height.

Three control parameters: the tensile strength f_t, the maximum cracking opening width w_c and the fracture energy G_F, of the softening curves obtained through inverse analysis from three wedge splitting specimens WS13~WS16 (d_{max}=20mm, design strength: C20), WS22~25 (d_{max}=80mm, design strength: C20, dam concrete) and WS32~35 (d_{max}=40mm, design strength C20, wet-screening concrete) were compared as shown in Fig. 10-27 through Fig. 10-29.

Fig. 10-29 shows that with the increase of ligament height H, the fracture energy G_F increases then tends to be stable, which shows the size effect. This may be attributed to a multitude of causes. For instance, the specimen placement, data log and many factors during the experiment may affect G_F. Although some energy dissipation may impact G_F, it is not the main cause to the size effect of G_F, after correcting these dissipations, G_F still shows the size effect.

10.7 Softening Behaviors of Dam Concrete and Wet-screening Concrete

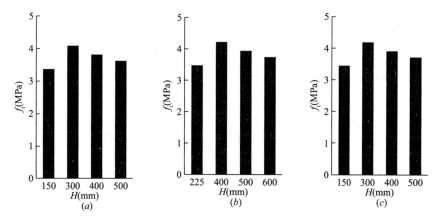

Fig. 10-27　Effect of ligament height H on f_t

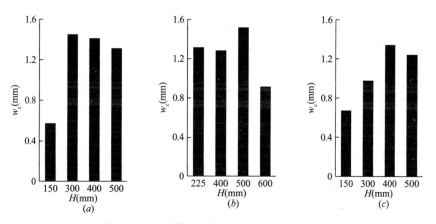

Fig. 10-28　Effect of ligament height H on w_c

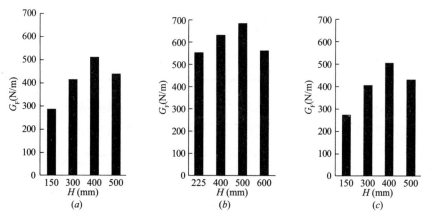

Fig. 10-29　Effect of ligament height H on G_F

After taking various factors into account, the author believes that the size effect of G_F identified herein may be reasonably explained as: the impact of boundary effect. To make it clear, the local fracture energy g_f should be introduced into the virtual crack model, as shown in Fig. 10-30. g_f refers to a local fracture energy on the new crack surface immediately after the fracture process zone (FPZ). According to the virtual crack model, the equation can be obtained as:

$$G_F = g_f = \int_0^{w_c} \sigma dw \qquad (10\text{-}11)$$

Where σ is the cohesive stress; w is the cracking opening width. That is to say, the local fracture energy g_f is the energy consumed when the crack tip zone w increases from 0 to w_c. Therefore, according to the virtual crack model, g_f should be maintained as a constant with the crack development.

However, the boundary effect arose from the cracking process of wedge splitting specimens, as shown in Fig. 10-31. That is to say, g_f will gradually decrease within the bottom boundary zone of specimen, and the fracture energy G_F is not evenly distributed on the entire fracture face (ligament). Therefore, the G_F measured showed the size effect. Some scholar proposed a bilinear model for the local fracture energy g_f.

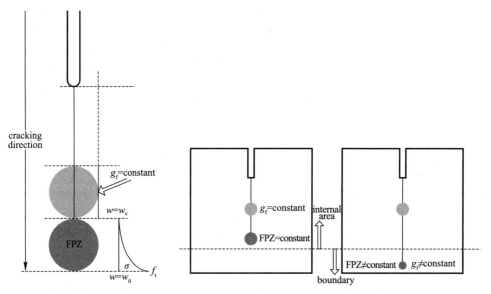

Fig. 10-30 Local fracture energy g_f Fig. 10-31 Boundary effect

This model shows that the ratio of the boundary zone to the whole specimen size becomes smaller, and the impact of boundary effect on fracture energy will also gradually diminish, with the increase of specimen size. Therefore, with the increase of specimen size, the fracture energy G_F will increase then tends to be stable. This is consistent with these findings.

As shown in Figs. 10-27 and 10-28, the tensile strength f_t does not change greatly with the increase of specimen size. The change trend of the maximum cracking opening width w_c is similar to the fracture energy G_F, both of them increase then tend to be stable.

Fig. 10-32 shows that relationship between the softening curve and the fracture process zone (FPZ) obtained from the literature.[27] The softening curve can be simply divided into two parts. The first part (including f_t) is mainly associated with the localization and propagation of micro-cracks, and the process is not substantially affected by the specimen size. The second part (including w_c) is associated with the visible cracks developed from micro-cracks through bridging effect, since the first half of wedge splitting specimens with different sizes is similar, and the fracture energy G_F shows the size effect, the maximum cracking opening width w_c shows the similar trend with G_F.

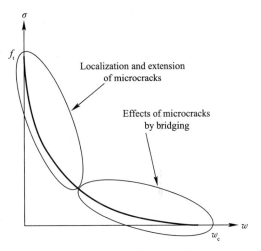

Fig. 10-32 Relationship between softening curve and fracture process zone

10.7.4 Comparison of softening curves between dam concrete and wet-screening concrete

The appropriate determination of mass concrete mechanical performance indicators is critical to the design and construction of safe, reliable and economical structures. The mechanical performance test of fully-graded mass concrete is often affected by the specimen size, testing machine, testing method and other factors, which is very difficult to perform. Previously, the wet-screening concrete specimens (the coarse aggregates larger than 40mm were screened out of the mixed three or four graded dam concrete using the 40mm sieve, the concrete after wet screening was prepared into standard specimens) were widely used in the domestic and foreign experiments, and the mechanical performance indicators of wet-screening concrete were used as the mechanical performance indicators for fully-graded concrete. However, it is necessary to determine the relationship due to the differences between them. The performance indicators of wet-screening concrete were used to deduce the performance indicators of fully-graded mass concrete, thereby simplifying the test procedures.

The wedge splitting specimens used herein had three different sizes. The wet-screening concrete specimens WS33~WS35 corresponded to the prototype grade dam concrete WS23~WS25 with the maximum aggregate size of 80mm. For the convenience of drawing, these three sizes were referred to as one, two and three.

Three control parameters: f_t, w_c and G_F of the softening curve obtained through inverse analysis from the above dam concrete and wet-screening concrete , as shown in Fig. 10-33 .

Fig. 10-33(a) shows that f_t of fully graded dam concrete was less than that of the corresponding wet-screening concrete, and with the increase of specimen size, their differences increased then tends to be stable. It maybe result from the follows: Larger size aggregate made concrete more inhomogeneous and more weak so that the tensile strength f_t of fully graded dam concrete was less than that of wet-screening concrete. However, with the further increase of specimen size, the tendency as the "large aggregate" for 80mm aggregate concrete tends to weaken. It led their differences increased then tended to be stable.

Fig. 10-33(c) shows that G_F of fully graded dam concrete was larger than that of the corresponding wet-screening concrete. With the increase of specimen size, the differences between them increased and then tended to be stable. The effect was similar to the effect of aggregate on tensile strength f_t above-mentioned. In addition, because the size of dam concrete specimen is larger than that of the wet-screening concrete specimen, the impact of size effect offset a partial effect of aggregate size, but it was not enough to eliminate the differences between dam and wet-screening concrete. As shown in Fig. 10-33 (b), the difference of w_c between the fully graded dam concrete and the wet-screeening concrete may be attributed to the mutual influence of the G_F increasing and the f_t decreasing.

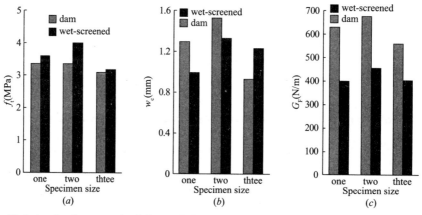

Fig. 10-33 Relationship between the fully-graded dam concrete and corresponding wet-screening concrete

10.8 Conclusions

This research performed the wedge splitting tests on 19 groups of dam concrete and wet-screening concrete specimens (76 in total) and the direct tension tests on 8 groups of the corresponding materials (24 specimens in total). The results can be obtained as follows:

(1) According to the principle of fracture mechanics of concrete, the expressions of softening curve of dam concrete and wet-screening concrete were achieved by the direct tension method.

(2) As the test data of a wedge splitting specimen contained nearly hundreds of thousands lines data, the processing method of *P-CMOD* curve to remove some inappropriate data points was proposed and used for the subsequent inverse analysis calculation. It didn't perform the inverse analysis calculation after averaging the *P-CMOD* curves of specimens in the same group, but solved the softening curve for each specimen.

(3) The FEM procedure for calculating softening curves of dam concrete and wet-srceening concrete by inverse analysis was developed. It introduced the modified optimization algorithm based on neighborhood attraction into the inverse analysis calculation, reduced risk of getting into

the local minimum, and improved calculation efficiency of inverse analysis.

(4) The average fitting method of softening curves of a specimen group was proposed which can better yield a softening curve of concrete that represents a group of wedge splitting specimens.

(5) When the specimen aggregate size is less than 40mm, the softening curve obtained from inverse analysis and direct tension method was basically consistent, which verified the feasibility of inverse analysis for determining the softening curve. For the fully-graded dam concrete, the softening curve determined by these two methods was different, which needs to be further examined.

(6) With increase of concrete compressive strength, the tensile strength f_t, the fracture energy G_F and the maximum cracking opening width w_c obtained from the inverse analysis increased then tended to be stable.

(7) For wedge splitting specimens with the same specimen size and strength, the maximum cracking opening width w_c and fracture energy G_F obtained by inverse analysis increased with the increase of aggregate size, while the tensile strength f_t increased and then decreased with the increase of aggregate size.

(8) The G_F of dam concrete and wet-screening concrete obtained by inverse analysis existed size effect. With the increase of specimen size, G_F increased and then tends to be stable, which showed the boundary effect. The f_t changed a little and w_c showed the size effect.

(9) The f_t of fully-graded dam concrete obtained by the inverse analysis was less than that of the corresponding wet-screening concrete, and for the G_F the situation was opposite.

References

[1] Gopalaratnum, V.S. and Shah, S.P. . Softening response of plain concrete in direct tension. ACI Materials Journal, 1985, 82(3): 310-323.

[2] Phillips, D. and Zhang, B. . Direct tension tests on notched and unnotched plain concrete specimens, Magazine of Concrete Research, 1993, 145(162): 25-32.

[3] Li, Q.B. , Ansari, F. . High strength concrete in uniaxial tension. ACI Materials Journal, 2000, 97(1): 49-57.

[4] Roelfstra, P.E. and Wittmann, F.H. . Numerical method to link strain softening with failure of concrete, Fracture Toughness and Fracture Energy of Concrete, F.H. Wittmann, ed., Amsterdam, Elsevier, 1986, pp. 163-175.

[5] Kitsutaka, Y. . Fracture Parameters by Polylinear Tension-Softening Analysis, Journal of Engineering Mechanics, 1997, 123(5): 444-450.

[6] Slowik, V., Villmann, B., Bretschneider, N. and Villmann, T. . Computational aspects of inverse analyses for determining softening curves of concrete. Computational Methods in Applied Mechanics and Engineering, 2006, 195: 7223-7236.

[7] Kwon, S.H., Zhao, Z.F. and Shah, S.P. . Effect of specimen size on fracture energy and softening curve of concrete. Part II : Inverse analysis and softening curve. Cement and Concrete Research, 2008, 38(8-9): 1061-1069 .

[8] Villmann, T. . Evolutionary algorithms with subpopulations using a neural network like migration scheme

and its application to real world problems [J]. Integr. Comput.-Aided Engrg. 2002, 9: 25-35.

[9] Cai, Y.B., Li, J.Y. . Test code for hydraulic concrete (DL/T 5150-2001) [S]. BeiJing: China WaterPower Press, 2006.

[10] Yang, L.J. . Research on softening relationships of dam concrete based on the wedge splitting tests [D]. Dissertation of Zhejiang University of Technology, 2010.

[11] Yang, W.N. .The Quadratic Spline Approximation of Curvature Interpolation and Its Convergence [J]. Journal of Chongqing University, 1982, 4: 79-85.

[12] Huang, M.L., Wang, X.D. and Cao, L. . Effect of support forms on fracture toughness KIC of wedge splitting specimens [J]. Journal of Hohai University(Natural Sciences) , 2006, 34 (4): 435-439.

[13] CEB-FIP Model Code 1990: Design Code [S]. 34-38.

[14] Gopalaratnama, V. S. and Ye, B. S. . Numerical characterization of the nonlinear fracture process in concrete [J]. Engineering Fracture Mechanics, 1991, 40 (6): 991-1006.

[15] Japan Concrete Institute, Determination of tension softening diagram of concrete, JCI-TC992, Test Method for Fracture Property of Concrete, Draft, 2001.

[16] Moscato, P. . On evolution, search, optimization, genetic algorithms and material art: towards memetic algorithms, C3P 826, California Institute of Technology, 1989.

[17] Huhse, J., Villmann, T., Merz, P., Zell, A. . Evolution strategy with neighborhood attraction using a neural gas approach [J]. Parallel Problem Solving from Nature VII, Lecture Notes in Computer Science, 2002, 2439: 391-400.

[18] Huhse, J., Villmann, T. and Zell, A. . Investigation of the neighborhood attraction evolutionary algorithm based on neural gas [J]. Neural Networks and Soft Computing, 2003, 340-345.

[19] Villmann, T., Villmann, B. and Slowik, V. . Evolutionary algorithms with neighborhood cooperativeness according neural maps [J]. Neurocomputing, 2004, 57: 151-169.

[20] Huang, M.L. , Wang, X.D. and Cao, L. . Effect of support forms on fracture toughness K_{IC} of wedge splitting specimens [J]. Journal of Hohai University (Natural Sciences) , 2006, 34 (4): 435-439.

[21] Jia, Y.D. . Different coarse aggregate and strength of concrete fracture performance and test method research [D]. Dissertation of Dalian University of Technology, 2003.

[22] Hou, X.Y . Effect of coarse aggregate on mechanical properties of concrete discussions [J]. Water Conservancy Science and Technology and Economy, 2011, 10 (17): 100-102.

[23] Zhang, J., Leung, C.K.Y., Xu, S.L. . Evaluation of fracture parameters of concrete from bending test using inverse analysis approach [J]. Materials and Structures, 2010, 6 (43):857-874.

[24] Qu, H.C., Chen, G.R. and Xia, X.Z. . Influence of aggregate shape on tensile strength of concrete [J]. Journal of hohai university (natural science), 2008, 36 (4): 554-558.

[25] Bazant, Z.P. and Chen, E.P.. Scaling law of Structural damage [J], Advances in Mechanics, 1999, 29 (3): 383-433.

[26] Hillerborg, A.. Results of three comparative test series for determining the fracture energy G_F of concrete [J]. Materials and Structures, 1985, 18(5): 407-413.

[27] Nomura, N., Mihashi, H. and Izumi, M. . Correlation of fracture process zone and tension softening behavior in concrete [J]. Cement and Concrete Research, 1991, 21(4): 545-550.

11 Effect of Processing of Tail Section of Tested Curve on Fracture Energy of Dam Concrete

11.1 Introduction

The fracture energy of dam concrete is a very important property which can characterize fracture of dam concrete. The coarse aggregate size of dam concrete is 80mm in this paper. The wet-screening procedure is to remove all aggregate particles large than 40mm from the fresh dam concrete. Wet-screened concrete specimens are widely adopted to carry out experiments to evaluate the physical and mechanical properties of dam concrete. This is an approximate method.

The wedge-splitting tests were performed on the specimens which the mix were designed for an actual dam built in China to experimentally investigate the fracture energy of dam concrete and wetscreened concrete. The 12 specimens were divided into 3 groups with different specimen sizes for dam concrete, and another 16 specimens were divided into 4 groups for wet-screening concrete. The details of experiments see our research report (Zhao 2004, Zhou et al. 2004).

Based on the principle of work-of-fracture (RILEM 1985), the fracture energy G_F of dam concrete and wetscreened concrete were calculated by the tested *P-CMOD* curve. It was shown that the stable tested *P-CMOD* curves of dam concrete and wet-screening concrete can be obtained by employing the testing method in this paper. However, the measured maximum value of CMOD of some specimens exceeded capacity of the displacement sensor which is 5mm due to employing large size and large aggregate dam concrete specimens in the fracture test. It led to CMOD values of the tail section of the tested *P-CMOD* curves of some specimens distorted. The tail section of the tested *P-CMOD* curve need to be fitted, so that the complete *P-CMOD* curve can be got for calculating the G_F of wedge-splitting specimen.

11.2 Determination of Fracture Energy of Dam Concrete and Wet-screening Concrete by Wedge-splitting Test

The test data come from our research report (Zhao 2004, Zhou et al. 2004), including 3 different sizes wedge-splitting dam concrete specimens and 4 sizes wet-screening concrete specimens. The geometry of specimen is shown in Fig. 11-1 and sizes are shown in Table 11-1.

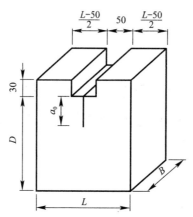

Fig. 11-1 Geometry of wedge-splitting specimen (unit:mm)

Sizes of wedge-splitting specimens   Table 11-1

Specimen	Maximum aggregate size (mm)	Sizes of specimens (mm)			
		B	D	L	a_0
WS-d-1	80	250	450	450	225
WS-d-2			800	800	400
WS-d-3			1000	1000	500
WS-w-1	40	200	300	300	150
WS-w-2			600	600	300
WS-w-3			800	800	400
WS-w-4			1000	1000	500

There were 4 companion specimens in each specimen group. By wedge-splitting test, the P-CMOD curve of 3 companion specimens for each group can be tested successfully.

11.2.1 Processing of tail section of tested P-CMOD curve for wedge-splitting specimen

Take a specimen as an example to explain how to obtain a complete P-CMOD curve by processing tail section of the tested curve, then to determine the G_F of dam concrete and wet-screening concrete.

Some specimens may have the almost complete tested P-CMOD curves by wedge-splitting test. For example the specimen WS-1 (see Fig. 11-2(a)), the load of tail section of the tested P-CMOD curve reaches to P=80N, compared with the peak load P_{max}=4594.8N. It can be considered that the load value closes to zero. The rationality of fitting the tail section of the tested curve by power function or exponential function was investigated by contrasting between tail section of completely tested P-CMOD curve and that of the fitted curve.

In order to got the area enclosed by a complete P-CMOD curve, fitting tail section of the tested curve is needed. According to reference (Xu S.L. et al. 2007), the part after inflection point of descending part of the tested P-CMOD curve(about at the 1/3 of peak load of tail section) was

11.2 Determination of Fracture Energy of Dam Concrete and Wet-screened Concrete by Wedge-splitting Test

fitted by by power function and exponential function.

In wedge-splitting test, P_v is the vertical load and $CMOD$ is the crack opening displacement. The fitting formula by power function is:

$$P_v = \beta(CMOD)^{-\lambda} \quad (\beta, \lambda > 0) \tag{11-1}$$

The fitting formula by exponential function is:

$$P_v = me^{-n(CMOD)} \tag{11-2}$$

The fitting process of the tail section of tested P-$CMOD$ curve by power function is as follows: The tested P-$CMOD$ curve is shown in Fig. 11-2(a). The part of tested curve from inflection point to tail section which need to be fitted is shown in Fig. 11-2(b). Fitted the curve by formula (1) whose correlation coefficient is $R^2 = 0.9981$, the fitting result is shown in Fig. 11-2(c). The Fig. 11-2 shows that the fitting tail section is a good agreement with the corresponding tested curve. The fitting tail section curve by power function descends slowly than the corresponding tested curve.

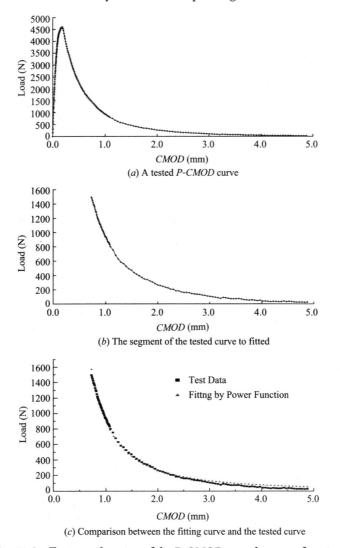

(a) A tested P-$CMOD$ curve

(b) The segment of the tested curve to fitted

(c) Comparison between the fitting curve and the tested curve

Fig. 11-2 Fitting tail section of the P-$CMOD$ curve by power function

The fitting process of tail section of the tested curve by exponential function of wedge-splitting specimen is just like by power function, but with the formula (11-2) in curve fitting, the fitting result is shown in Fig. 11-3. The area enclosed by the fitting tail section curve by exponential function is some less than that by the tested curve and the fitting tail section curve descends rapidly than the corresponding tested curve.

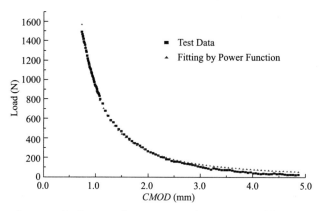

Fig. 11-3 Comparison between the fitting tail section curve by exponential function and the corresponding tested curve

11.2.2 Calculation of fracture energy of dam concrete and wet-screened concrete of wedge-splitting specimen

A *P-CMOD* curve is an important tested curve obtained by wedge-splitting test. For each of 3 companion specimens tested in a specimen group, the original test data of the *P-CMOD* curve reached up to tens of thousands rows to hundreds of thousands rows. The original test data were processed by the procedure such as filtering the test data scattered far from the *P-CMOD* curve, etc. Then the *P-CMOD* curve of 3 companion specimens for each group were averaged based on crack propagation process of concrete materials. Thus a representative *P-CMOD* curve (or called averaged *P-CMOD* curve) which can characterize the fracture characteristics of the companion specimens can be obtained (Zhao Z.F. et al. 2009).

The following three methods were employed to calculate the fracture energy G_F of dam concrete and wet-screening concrete: ① Based on the principle of work-of-fracture, calculated the G_F by the tested *P-CMOD* curve for each companion specimens, then averaged the results and got the G_F of each specimen group. ② Calculated the G_F by the averaged *P-CMOD* curve of which the tail section was fitted by the power function. ③ Calculated the G_F by the averaged *P-CMOD* curve of which the tail section was fitted by the exponential function. Then compare the different G_F. The G_F obtained by wedgesplitting test can be calculated as follows:

$$G_F = W_0/A_0 \tag{11-3}$$

$$W_0 = W/2\tan\theta \tag{11-4}$$

11.2 Determination of Fracture Energy of Dam Concrete and Wet-screened Concrete by Wedge-splitting Test

A_0 is ligament area. $W_0 = W/2\tan\theta$. θ is the wedge angle of the test set-up. As shown in Fig. 11-4, W is the area under P-$CMOD$ curve in the ①method. In the ②method and ③method, $W = W_1 + W_2$. W_2 is the area under the part of the tested P-$CMOD$ curve before the fitted tail section of curve. W_1 is the area under the fitted tail section curve. The formula for calculating W_1 by power function is as equation (11-5), while by exponential function is as equation (11-6):

$$W_1 = \int_{CMOD_1}^{\infty} \beta(CMOD)^{-\lambda} \, d(CMOD) = \frac{\beta}{(\lambda-1)(CMOD_1)^{(\lambda-1)}} \tag{11-5}$$

$$W_1 = \int_{CMOD_1}^{\infty} me^{-n(CMOD)} \, d\delta = \frac{m}{ne^{nCMOD_1}} \tag{11-6}$$

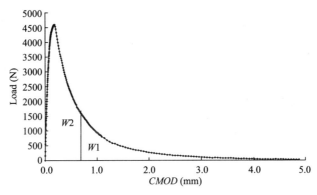

Fig. 11-4 P-$CMOD$ curve by wedge-splitting test

11.2.3 Comparative analysis of the fracture energy calculated by different methods

The fracture energy G_F calculated by the above three methods are shown in Table 11-2. According to Table 11-2, the G_F calculated by method ①＜the G_F calculated by method ②＜the G_F calculated by method ③.

Comparison of the G_F obtained by different methods   Table 11-2

Specimon	Category	D (mm)	G_F by Method① (N/m)	G_F by Method② (N/m)	G_F by Method③ (N/m)
WS-d-1	Dam concrete	450	362.1	385.5	514.2
WS-d-2		800	598.1	666.3	819.9
WS-d-3		1000	516.8	558.2	633.8
WS-w-1	Wet-screening concrete	300	203.2	218.6	243.7
WS-w-2		600	354.1	373.3	439.8
WS-w-3		800	484.1	523.1	667.0
WS-w-4		1000	369.4	423.6	562.1

The G_F calculated by method ①was the smallest because of not considering the influence of tail section curve. Compared with the peak load between 5.03kN to 24.21kN, the end point of tested load existed between 0.18kN to1.89kN which can be regarded approaching to zero load. Therefore the obtained G_F was a good agreement with the actual fracture energy. The G_F calculated by method ③was overestimated because two reasons: one was the tail section curve fitted by power function descended slowly than the corresponding tested curve, another is the *CMOD* of fitted tail section curve approaching infinity compared with the tested maximum *CMOD* used in method ①. The tail section curve fitted by exponential function descended rapidly than the corresponding tested curve, which led to the calculated G_F being smaller. However, the *CMOD* of fitted tail section curve approaching infinity compared with the tested maximum *CMOD* led to the calculated G_F overestimated. In sum, the G_F calculated by method ②is possibly the nearest to that calculated by the corresponding tested *P-CMOD* curve. The results in Table 11-2 confirmed this point. So the G_F of dam concrete and wet-screening concrete were identified by the averaged *P-CMOD* curve of which the tail section curve was fitted by the exponential function for each specimen group.

11.3 Discussion of the Fracture Energy of Dam Concrete and Wet-screened Concrete

The G_F of different sizes specimens of dam concrete and wet-screening concrete were shown in Fig. 11-5. It can be seen that G_F of these two kinds of concrete increased with an increase of the specimen size and the asymptotic behavior over the size was found. It also showed that G_F of dam concrete was larger than that of wet-screening concrete over the same specimen size.

Recently, some scholars (Duan K. et al. 2003) investigated the size effect of G_F from the boundary conditions of specimen. The boundary effect on fracture energy of dam concrete and wet-screening concrete would be tried in further study.

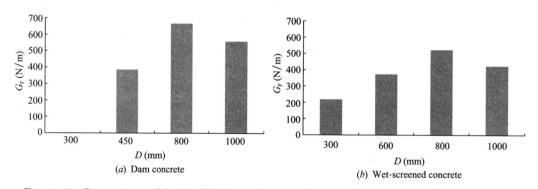

Fig. 11-5 Comparison of the G_F of different sizes specimens of dam concrete and wet-screening concrete

11.4 Conclusions

The measured maximum *CMOD* value of some specimens which were included in our 7 groups wedge-splitting specimens with different sizes for dam concrete and wet-screening concrete exceeded capacity of the displacement sensor (5mm). It led to *CMOD* values distorted at the tail section of the tested *P-CMOD* curves for some specimens. The tail section curve need to be fitted, so that a complete *P-CMOD* curve can be got for calculating G_F.

The comparison between the fitted tail section curve by power function or exponential function and the corresponding tested complete curve showed that the fitted tail section curves by these two functions were a good agreement with the corresponding tested curves. The area enclosed by the fitted tail section curve by power function was some larger than that by the tested curve and the fitted tail section curve descended slowly than the corresponding tested curve. The area enclosed by the fitted tail section curve by exponential function was less than that by the tested curve and the fitted tail section curve descended rapidly than the corresponding tested curve.

Three methods were employed to calculate the fracture energy of dam concrete and wet-screening concrete: ①Calculated G_F by the tested *P-CMOD* curve for each companion specimen, then averaged the results and got the G_F for each specimen group. ②Calculated G_F by the averaged *P-CMOD* curve of which the tail section was fitted by exponential function. ③Calculated G_F by the averaged *P-CMOD* curve of which the tail section was fitted by power function. It was shown that the G_F obtained by method ①was slightly small, but mostly close to the actual value. The G_F obtained by method ③was some larger. The G_F obtained by method ② was the nearest to the result calculated by the originally tested complete *P-CMOD* curve. So the fracture energy of dam concrete and wet-screening concrete were identified by the method ②

It was shown that G_F of dam concrete and wetscreened increased with an increase of the specimen size and the asymptotic behavior over the size was found. It was also shown that G_F of dam concrete was larger than that of wet-screening concrete over the same specimen size. The boundary effect on fracture energy of dam concrete and wet-screening concrete would be tried in further study.

References

[1] Duan, K., Hu, X.Z. & Wittmann, F.H. 2003. Boundary effect on concrete fracture and non-constant fracture energy distribution. *Engineering Fracture Mechanics*. 70: 2257-2268

[2] RILEM Draft Recommandation(50-FCM). 1985. Determination of the fracture energy of mortar and concrete by means of three-point bend tests on notched beams. *Materials and Structures*. 23 (138): 457-460

[3] Xu, S.L., Bu, D. & Zhang X.F. 2007. Determination of fracture energy of concrete using wedge-splitting test on compact tension specimens. *Journal of Hydraulic Engineering*. 38 (6): 683-689

[4] Zhao, Z.F. 2004. Research on fracture behaviors of dam concrete based on cohesive force of cracks. *Postdoctoral Research Report of Tsinghua University and China Gezhouba (Group) Corporation.*

[5] Zhao Z.F., Yang L.J., Zhao Z.G. & Zhu M.M. 2009. Test Data Processing Method of Fracture Experiments of Dam Concrete for Inverse Analysis. In Yuan, Y., Cui, J.Z. & Mang,H.A. (eds). *Proceedings of the International Symposium on Computational Structural Engineering, Shanghai, 22-24 June* 2009. Springer.

[6] Zhou, H.G., Li, Q.B. & Zhao Z.F. 2004. Research on fracture simulation of mass concrete cracks. *Cooperative Research Report of China Gezhouba(Group) Corporation, Tsinghua University and Yantai University.*

12 Experimental Study for Determining Double-K Fracture Parameters of the Three Gorges Dam Concrete

12.1 Introduction

The total cast quantity of the unparalleled China's Three Gorges Dam key water control project is 27.94 million cubic meters. It's fairly important for the dam designing, constructing, predicting and evaluating of the dam security and durability to study the dam concrete fracture features.[1] Tests of the fracture features of dam concrete are usually influenced by specimen sizes, testing machines and testing methods, so it's very difficult.[2] Wet-screening concrete specimens are widely adopted at home and abroad to carry out experiments, and regard the performance indexes of them as those of dam concrete. However, these two could not be completely identical, we should ascertain the quantitative relation between them and can determine the dam concrete performance indexes by using those of wet-screening concrete. Thus, the dam can be designed and constructed more scientifically and rationally. To obtain the quantitative relationship between the wet-screening concrete and the dam concrete of the Three Gorges dam. This chapter obtained the softening constitutive relation between the two through uniaxial tension tests, according to the double-K fracture model theory proposed by Xu and Reinhardt [3-5], determining the relevant fracture parameters of the two and obtaining the relationship of initial fracture toughness and unstable fractures toughness of the two through there-point bending notched beam test. The test can provide experimental method and data for the edit of the national industry standard "hydraulic concrete fracture test procedures", which provides a basis for the analysis of simulative fracture of the discharged segment and provides test data and research ideas for the study of the fracture properties of the concrete of Three Gorges Dam.

12.2 Double-K Fracture Parameters Based on the Cohesive Stress

In the study of fracture mechanics of concrete, many scholars have put forward various fracture models, such as the fictitious crack model, crack band model, two parameters model, size effect model, the effective crack model, double-K fracture model and the resistance curve model and so on. Among them, double-K fracture model is the model founded on the cohesion of the fracture process zone, with linear elastic fracture mechanics being its' foundation. It takes advantage of the fictitious crack model and the revised linear elastic fracture mechanics model, combining the cohesion of the seam between the front of the fracture process zone with the stress intensity factor to quantitatively describe the whole process of the concrete crack propagation. According to the double-K fracture criterion[3-5],

$$\begin{cases} K_I < K_{IC}, & \text{no crack propagation} \\ K_{IC}^{ini} \leq K_I \leq K_{IC}^{un}, & \text{steady crack propagation} \\ K_I > K_{IC}^{un}, & \text{crack propagation out of balan} \end{cases}$$

Where, K_I is the stress intensity factor, K_{IC}^{ini} is the initiation fracture toughness, and K_{IC}^{un} is the unstable fracture toughness which are so-called the double-K fracture parameters.

By lend of the experimental results of the load-crack mouth opening displacement curve (P-CMOD curve) for a there-point bending notched beam under monotonic loading, and defining the softening constitutive relation σ-w curve, we can use the analytic method to calculate the double-K fracture parameters of the concrete materials.

12.2.1 Critical effective crack length

Fig. 12-1 is the schematic diagram of a there-point bending notched beam. According to the linear progressive superimposed assumptions, the process of the nonlinear progressive assumptions can be simplified as a series of linear superposition process. We can calculate the critical crack length a_c by substituting the maximum load measured P-CMOD curve P_{max} and the critical crack mouth opening displacement $CMOD_c$ values into the equation (12-1).

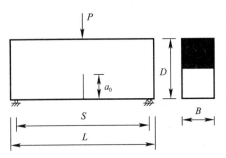

Fig. 12-1 Configuration of a there-point bending notched beam

$$CMOD_c = \frac{6P_{max}Sa_c}{D^2 BE}V_r(a) \qquad (12-1)$$

Where S represents beam span, D represents height, B represents thickness, E represents elastic modulus, The function $V_r(\alpha)$ to the standard there-point bending notched beam of the span-depth ratio $S/D=4$ is

$$V_r(\alpha) = 0.76 - 2.28\alpha + 3.87\alpha^2 - 2.04\alpha^3 + \frac{0.66}{(1-\alpha)^2} \qquad (12-2)$$

where, $\alpha = (a_c + H_0)/(D + H_0)$; H_0 represents thickness of clamp meter blade.

12.2.2 Cohesive toughness

According to equation (12-3), we can use the measured $CMOD$ to obtain the critical crack tip opening displacement $CTOD$

12.2 Double-K Fracture Parameters Based on the Cohesive Stress

$$CTOD_c = CMOD_c \left\{ \left(1 - \frac{a_0}{a_c}\right)^2 + \left(1.081 - 1.149\frac{a_c}{D}\right)\left[\frac{a_0}{a_c} - \left(\frac{a_0}{a_c}\right)^2\right] \right\}^{1/2} \quad (12\text{-}3)$$

In the critical state of crack propagation, the cohesive stress distribution is shown in Fig. 12-2

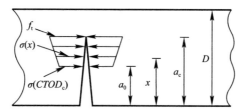

Fig. 12-2 Distribution of the cohesive force in the critical state

The expression of the cohesive stress is

$$\sigma(x) = \sigma(CTOD_c) + \frac{x - a_0}{a_c - a_0}[f_t - \sigma(CTOD_c)] \quad (12\text{-}4)$$

In the critical state, the cohesive toughness caused by cohesion $\sigma(x)$ is

$$K_{IC}^c = \int_{a_0}^{a_c} 2\frac{\sigma(x)}{\sqrt{\pi a_c}} F_1\left(\frac{x}{a_c}, \frac{a_c}{D}\right) dx \quad (12\text{-}5)$$

$$F_1\left(\frac{x}{a_c}, \frac{a_c}{D}\right) = \frac{3.52\left(1 - \frac{x}{a_c}\right)}{\left(1 - \frac{a_c}{D}\right)^{3/2}} - \frac{4.35 - 5.28\frac{x}{a_c}}{\left(1 - \frac{a_c}{D}\right)^{1/2}}$$

$$+ \left\{\frac{1.30 - 0.30\left(\frac{x}{a_c}\right)^{3/2}}{\sqrt{1 - \left(\frac{x}{a_c}\right)^2}} + 0.83 - 1.76\frac{x}{a_c}\right\}\left\{1 - \left(1 - \frac{x}{a_c}\right)\frac{a_c}{D}\right\} \quad (12\text{-}6)$$

12.2.3 Unstable fracture toughness

As for STPB, we can obtain the unstable fracture toughness by substituting the measured P_{max} and calculated a_c into the equation (12-7).

$$K_{IC}^{un} = \frac{3(P_{max} + p)S}{2D^2 B}\sqrt{a_c} F_3\left(\frac{a_c}{D}\right) \quad (12\text{-}7)$$

Where p represents the equivalent weight load of the beam.

$$F\left(\frac{a_c}{D}\right) = \frac{1.99 - \frac{a_c}{D}\left(1 - \frac{a_c}{D}\right)\left[2.15 - 3.93\frac{a_c}{D} + 2.7\left(\frac{a_c}{D}\right)^2\right]}{\left(1 + 2\frac{a_c}{D}\right)\left(1 - \frac{a_c}{D}\right)^{3/2}} \qquad (12\text{-}8)$$

12.2.4 Fracture toughness

Fracture toughness is equal to the difference between the unstable fracture toughness and the cohesive toughness, that is

$$K_{IC}^{ini} = K_{IC}^{un} - K_{IC}^{i} \qquad (12\text{-}9)$$

12.3 Specimen Design and Test Methods

12.3.1 Raw materials and mix propotion

Raw materials: the cement is the Three Gorges brand medium thermal Portland cement produced in Gezhouba cement plant, the fly ash is the one-grade fly ash produced by the thermal power plant in Zouxian of Shandong province, the coarse aggregate is the artificial gravel from Three Gorges old trees ridge artificial aggregate processing system, the maximum aggregate size is 80mm, and the fine aggregate is the artificial sand of the shore creek of the Three Gorges, mixing the drinking water of the Three Gorges.

The mix proportion was based on the construction mix proportion of dam concrete, as shown in Table 12-1.

Mix proportion of dam concrete (unit: kg/m³) Table 12-1

Unit water amount	Water-to-binder ratio	Sand ratio (%)	Ratio of fly ash (%)	Cement	Fly ash	Sand	Small gravel	Medium gravel	Large gravel	Water reducer	Entraining agent
102	0.45	30	30	159	68	625	374	374	748	1.362	0.0159

12.3.2 Specimen fabrication

Use the drop-mixer to mix concrete, and the hose vibrator to make it dense. Remove the mold of the Specimen after pouring for 24 hours, cover them with straw bags in the natural environment, and conserve by watering them for 28 days. The testing age is (130±10) days. Sizes of the uniaxial tension specimen are 250mm×250mm×500mm for the dam concrete specimen. The specimen of is 150mm×150mm×300mm for the wet-screening concrete specimen. We use the steel plate coated with lubricant oil in both sides to make the prefabricated seam of the there-point bending notched beam. After the concrete initializing for 3 hours, remove the steel sheet from the concrete. The shapes and sizes of the specimens refer to Fig. 12-1 and Table 12-2.

Sizes of the three-point bending notched beams   Table 12-2

Category	No.	Specimen sizes (mm)				L/B	a_0/B
		B	D	L	a_0		
Dan concrete	SL43	240	400	1700	160	4	0.4
	SL46		550	2300	220		
Wet-screening concrete	SL47	120	200	900	80		
	SL48		250	1100	100		

12.3.3 Test equipment

The uniaxial tension tests were conducted by employing the digital servo-hydraulic closed-loop testing machine Instron 8506 3000kN with stiffness of 6GN/m to do the uniaxial tension test of concrete. The there-point bending notched beam test is experimented on the 5MN pressure testing machine controlled and loaded by the displacement.

12.4 Experimental Results and Analysis

12.4.1 Uniaxial tension test

Through the uniaxial tension test, we can get the relational expression of softening constitutive relation σ-w by measuring its' tensile strength f_t maximum crack width w_0 and fracture energy G_F. The tensile strength of dam concrete is f_t=2.052MPa, which value is less than that of wet-screening concrete f_t=2.601MPa. The fracture energy of dam concrete is G_F=232.778N/m, which is less than that of wet-screening concrete G_F=337.943N/m and is 68.9% of it. The crack width of dam concrete is w_0=0.73mm, which is more than that of wet-screening concrete w_0=0.71mm and is 1.03 times of it. The softening relation is as follows:

$$\sigma = f_t \left\{ 1 - \varphi \exp\left[-\left(\frac{\lambda}{w/w_0} \right)^n \right] \right\} \quad (12\text{-}10)$$

Where, φ, λ and n are material parameters. Using the method of least squares, we can obtain that: (1) In the dam concrete φ=1.21, λ=0.055, n=0.56, the related number of the theory and experiment is R^2=0.97. (2) In the wet-screening concrete φ=1.23, λ=0.068, n=0.55, the related number of the theory and experiment is R^2=0.98.

12.4.2 TPB test

The *P-CMOD* curves of part of the specimens are shown in Fig. 12-3.

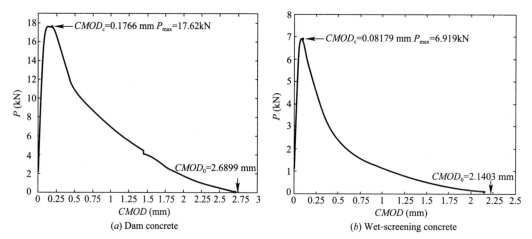

Fig. 12-3 **P-CMOD curve of part of the specimen of dam concrete and wet-screening concrete**

12.4.3 Determining of double-K fracture parameters

According to the σ-w relations of dam concrete and wet-screening concrete, combining with the there-point bending notched beam test, crack initiation fracture and unstable toughness of dam concrete and wet-screening concrete can be respectively obtained (see Table 12-4).

According to the study of Zhao and Xu, when the beam height $D \geqslant 200$mm, double-K Fracture Parameters of the there-point bending notched beam have no size effect, which can be confirmed in Table 12-4. According to the calculated result of the double-K Fracture Parameters, the crack initiation fracture toughness of dam concrete is 0.9193MPa·m$^{1/2}$, and its unstable toughness is 2.0349MPa·m$^{1/2}$, while the crack initiation fracture toughness and unstable toughness of wet-screening concrete are respectively 0.8070MPa·m$^{1/2}$ and 1.7065MPa·m$^{1/2}$. As we can see, both the initiation fracture toughness and unstable toughness of dam concrete are bigger than those of wet-screening concrete. The initiation fracture toughness of dam concrete is 1.14 times of that of wet-screening concrete, while the unstable toughness is 1.19 times of that of wet-screening concrete.

Measured specimen sizes P_{max} and $CMOD_c$ Table 12-3

No.	L(mm)	D(mm)	B(mm)	a_0(mm)	S(mm)	P_{max}(kN)	$CMOD_c$(μm)
SL43-1	1700	402	240	156	1600	17.608	177.84
SL43-2	1700	400	238	162	1600	17.635	124.55
SL43-3	1698	400	240	156	1600	23.406	130.75

continue

No.	L(mm)	D(mm)	B(mm)	a_0(mm)	S(mm)	P_{max}(kN)	$CMOD_c$(μm)
SL46-1	2299	553	240	222	2200	23.153	166.07
SL46-2	2299	550	239	223	2200	27.141	185.27
SL46-4	2298	550	240	218	2200	23.768	116.49
SL47-1	898	199	120	77	800	6.900	81.79
SL47-2	900	201	120	78	800	7.710	81.17
SL47-3	901	202	121	76	800	7.529	81.79
SL47-4	898	201	120	79	800	6.806	89.23
SL48-1	1091	249	121	98	1000	8.501	62.58
SL48-2	1098	250	120	96	1000	10.360	90.47
SL48-3	1097	249	120	94	1000	7.336	74.36
SL48-4	1096	251	120	94	1000	9.056	81.17

The double-K fracture parameters of dam and wet-screening concrete   Table 12-4

Dam concrete				Wet-screening concrete			
No.	D(mm)	K_{IC}^{ini} (MPa·m$^{1/2}$)	K_{IC}^{ini} (MPa·m$^{1/2}$)	No.	D(mm)	K_{IC}^{ini} (MPa·m$^{1/2}$)	K_{IC}^{ini} (MPa·m$^{1/2}$)
SL43-1	402	0.7491	2.2182	SL47-1	199	0.7478	1.7230
SL43-2	400	0.7466	1.7954	SL47-2	201	0.8949	1.7722
SL43-3	400	1.1186	2.0206	SL47-3	202	0.7793	1.7524
				SL47-4	201	0.7429	1.7903
SL46-1	553	0.8616	2.0980	SL48-1	249	0.8146	1.4503
SL46-2	550	1.2260	2.3549	SL48-2	250	1.1266	1.9318
SL46-4	550	0.8139	1.7220	SL48-3	249	0.5227	1.5165
				SL48-4	251	0.8268	1.7158
Average value μ		0.9193	2.0349	average value μ		0.8070	0.7065
Standard deviation		0.2035	0.2431	Standard deviation		0.1688	0.1541
Coefficient of variation γ		0.2213	0.1195	Coefficient of variation γ		0.2092	0.0903

12.5 Conclusions

It is quite important for scientific design, construction, security prediction of the high dam to study the fracture features of the Three Gorges dam concrete. This chapter is about the study of the fracture Parameters of the dam concrete and wet-screening concrete. From the uniaxial tension test and the there-point bending notched beam test, adopting the calculation methods of the double-K fracture parameters based on the cohesive stress, the double-K fracture parameters of the dam concrete and wet-screening concrete can be determined. The results show that the initiation fracture toughness and unstable fracture toughness of the dam concrete are bigger than those of the wet-screening concrete, that is, 1.14 times for the initiation fracture toughness and 1.19 times for the unstable fracture toughness.

References

[1] ZHAO Zhi-fang. Research on fracture behaviors of dam concrete based on the crack cohesive force [D]. Postdoctoral Research Report of Tsinghua University and China Gezhouba Group company, Beijing, Tsinghua University 2004 (in Chinese).

[2] LI Qing-bin, DUAN Yun-ling, WANG Guang-lun. Behavior of large concrete specimen in uniaxial tension [J]. Magazine of Concrete Research, 2002, 54 (5): 385-391

[3] XU Shi-lang and Reinhardt H W. Determination of double-K criterion for crack propagation in quasi-brittle fracture. Part I Experimental investigation of crack propagation [J]. International Journal of Fracture, 1999, 98(2): 111-149

[4] XU Shi-liang and Reinhardt H W. Determination of double-K criterion for crack propagation in quasi-brittle fracture. Part II Analytically evaluating and practical measuring methods for there-point bending notched beams [J]. International Journal of Fracture, 1999, 98(2): 151-177

[5] XU Shi-liang and Reinhardt H W. Determination of double-K criterion for crack propagation in quasi-brittle fracture. Part III Compact tension specimens and wedge splitting specimens [J]. International Journal of Fracture, 1999, 98(2): 179-193

[6] ZHAO Zhi-fang and XU Shi-liang. Influence of specimen size upon new K_R resistance curves of concrete [J]. Journal of Hydropower Engineering, 2001, 12: 48-55 (in Chinese).

13 Experimental Research on Double-K Fracture Parameters of Dam Concrete with Various Aggregate Gradation

13.1 Introduction

The fracture mechanics of concrete is a effective tool to investigate cracks of concrete.

In recent forty years, major progress achieved in the field of concrete fracture mechanics can be summarized as follows: (1) Researchers used a variety of modern experimental observation techniques[1-3], found that there are stable extension phase of crack and the fracture process zone (FPZ) before unstable fracture of concrete crack. (2) The cohesive force exists in FPZ during crack propagation, which can be described by tensile softening constitutive relation quantitatively. (3) Prof. Hillerborg[4] proposed the Fictitious Crack Model (FCM) in which fracture energy express crack cohesive force in the FPZ. Prof. Bazant[5] proposed Crack Band Model (CBM). These crack models were successfully applied to model the failure process of concrete and reinforced concrete structures by FEM. (4) Some scholars proposed a variety of fracture models of concrete which consider stable extension of cracks before unstable fracture and take the critical stress intensity factor as concrete fracture model parameters, such as Two-Parameter Fracture Model by Jenq and Shah[6] Effect Crack Model, by Swartz[7] and Karihaloo[8], Size Effect Model by Bazant[9]. (5) In order to overcome deficiencies of the above-mentioned models, for example the fictitious crack model and crack band model were mainly employed in the finite element numerical analysis, Two-Parameter Fracture Model and Size Effect Model fails to consider the material cracks cohesive force and tensile softening characteristics as well as describe the whole process of concrete failure. The new fracture models which combine the crack cohesive force with equivalent crack and take the stress intensity factor as a parameter, such as double-K fracture model[10-11] proposed by Xu and Reinhardt and new K_R resistance curve based on cohesive force[12] were developed in recent years.

The related international academic organizations attached great importance to the standardization of determination of concrete fracture parameters based on the rich research achievements. For example, the RILEM have recommended a concrete type I fracture energy G_F standard testing method[13] in 1985 so that it lays the foundation for practical engineering application which employs FCM and CBM as the finite element structural analysis methods. Nordic 1999 "NT the BUILD 491" standard[14], which was based on the RILEM standard, recommended that the fracture energy of mortar and concrete for I mode crack can be determined by the tested load-deformation curve of a notched three-point bending beam (TPB) test. Unlike the above two criteria, the Japan Concrete Institute (JCI) proposed a fracture toughness of concrete test standard (JCI-TC992)[15] in 2001, which specified G_F determined by the tested load-crack mouth opening displacement (*P-CMOD*) curve of a TPB specimen. European model specification (CEB-FIP Model Code 90) has introduced I mode G_F and tension softening relation as the formal provision in early 1990s.[16] Since G_F can not be directly employed in analysis, further applications of concrete fracture mechanics method are subject to greater restrictions. For the structure and material design, the analysis method which takes stress intensity factor as the basic parameter is more convenient.

The double-K fracture model [10-12] which introduced two fracture toughness, initiation fracture toughness K_{IC}^{ini}, unstable fracture toughness K_{IC}^{um} to describe concrete crack propagation. The Double-K fracture model requires the less experimental parameters to be tested and the test procedure is simple. The Double-K fracture model was submitted to the U.S. ACI446 Commission as one of the five recommended draft concerning fracture parameters normalized testing method for discussion in October 2000. The further discussion was conducted at ASTM Symposium held on November 2001 in America. The Norm for Fracture Test of Hydraulic Concrete[17] took the Double-K fracture model as theoretical basis..

Under the guidance of academician Guofan Zhao who partially funded (purchasing the crucial testing equipment DH5937), Professor Hougui Zhou, Professor Qingbin Li and Professor Shilang Xu, China Gezhouba Group Company, Tsinghua University and Yantai University jointly accomplished the large-scale experiments charged by Prof. Zhifang Zhao and provided the test data. Dalian University of Technology and Gezhouba Group Company analyzed them. The initial cracking load rule and double-K fracture toughness and its size effect and shape effect of dam concrete with different gradation were analyzed and discussed based on the experiments which the largest specimens sizes were $S \times D \times B = 2200mm \times 550mm \times 240mm$ for TPB specimen and $2H \times D \times B = 1200mm \times 1200mm \times 250mm$ for WS specimen. It provided more basic test data for developing the Norm.

13.2 Experiments

13.2.1 Concrete mix proportions and basic mechanical properties

The concrete mix propotions is shown in Table 13-1. The cement employed is Jingmen medium heat 525 #. The fly ash is I-grade fly ash produced by Shandong Zhou County Power Plant. The coarse aggregate is artificial gravel obtained from Gushuling artificial aggregate processing system at Three gorges dam. Fine aggregate is artificial sand processed from Xiaanxi of the Three Gorges, mixed with drinking water of the Three Gorges. The relative density of cement, sand, gravel, fly ash were: $\gamma_{cement}=3.17$, $\gamma_{sand}=2.65$, $\gamma_{gravel}=2.72$, $\gamma_{fly\ ash}=2.14$. The Beijing Yejian JG3 superplasticizer was adopted. the dosage was 0.6%, and the solution concentration is 20%. The

DH9 produced by Shijiazhuang admixture factory was selected as air-entraining agent. The tested air content of concrete was 4.5%-5.5%. The age of specimens was approximately one year. Meanwhile the basic mechanical properties of the corresponding concrete were tested and the results are shown in Table 13-1 and Table 13-2.

13.2.2 Specimens

The standard three-point bending beam ($S/D=4$) and wedge-splitting specimen were employed. Aggregates include small size aggregates 10mm, 20mm and 40mm, and fully graded size aggregate 80mm. Concrete strength grade were C20, C30, C40, C50 and C25, respectively. The detailed specimen parameters are shown in Table 12-1, Table 12-2 and Table 12-3.

13.2.3 Arrangement of measuring points

The arrangement of measuring points of TPB specimen and WS specimen are shown in Chapter 1.2.

13.3 Determination of Initial Cracking Load P_{ini}

There exists a Fracture Process Zone (FPZ) and a certain amount of subcritical crack extension length for concrete fracture. Professor Shilang Xu[3-4] studied the whole fracture process of three-point bending beams and large compact tension specimens by laser speckle and photoelastic coating testing techniques etc, as shown in Fig. 13-1 and Fig. 13-2. The double-K fracture model was proposed based on observed results which show crack propagation process of concrete materials including three distinct stages: crack initiation, stable extension, unstable failure. The double-K fracture criterion can be simply stated as: the development of concrete crack is described by the initiation toughness and unstable fracture toughness. The crack occur initiation cracking when $K_I=K_{IC}^{ini}$. The crack propagates steadily when $K_{IC}^{ini}<K_I<K_{IC}^{un}$. The crack propagates unsteadily when $K_I>K_{IC}^{un}$. In practical applications, $K_I=K_{IC}^{ini}$ can be used as important structural crack propagation criterion; $K_{IC}^{ini}<K_I<K_{IC}^{un}$ can be used as a security alert before the important structural crack unstable propagation; $K_I=K_{IC}^{un}$ can be used as the criterion of the general structural crack propagation.

How to determine the cracking initiation point then to determine cracking initiation load P_{ini} is the key for calculating the initiation toughness of concrete. There are various methods for determining cracking load, such as laser speckle method, photoelastic coating method, acoustic emission method, the resistance strain gauge method, X-ray methods, scanning electron microscopy method and so on. Each of those methods has its own characteristics. The strain gauge method was employed in the tests of this research. The strain gauges were arranged on the both sides of the crack tip for observing the initiation of crack, also along the ligament for observing the propagating of crack. The details of arrangement of strain gauges see Chapter 1.2.

13 Experimental Research on Double-K Fracture Parameters of Dam Concrete with Various Aggregate Gradation

Table 13-1 Mix propotions of dam concrete

No.	Design label				Aggregate size (d_{max})	Grada-tion	Mix parameters					Material dosage per Cubic meter (kg)							
	Grade C	R	#	D	S			Water -bindt ratio	Unit water amount (kg/m³)	Fly ash content (%)	Sand ratio (%)	Ce-ment	Fly ash	Artificial sand	Small gravel	Medium gravel	Gravel Large gravel	Special large graver	Superplasti-cizer	Air-entrain-ing agent
1	20	90	200	250	10	10	One	0.50	140	30	45	196	84	869	1090	/	/	/	1.680	0.0196
2				250	10	20	One	0.50	132	30	43	185	79	846	1152	/	/	/	1.584	0.0185
3				250	10	40	Two	0.50	120	30	38	168	72	769	515	772	/	/	1.440	0.0168
4				250	10	80	Three	0.50	102	30	31	143	61	653	373	373	745	/	1.224	0.0143
5	25	90	250	250	10	80	Three	0.45	102	30	30	159	68	625	374	374	748	/	1.362	0.0159
6	30	90	300	250	10	20	One	0.45	135	20	42	240	60	814	1154	/	/	/	1.800	0.0195
7	40	90	400	250	10			0.35	135	20	40	309	77	744	1145	/	/	/	2.316	0.0251
8	50	90	500	250	10			0.30	140	10	39	420	47	698	1121	/	/	/	2.802	0.0280

Table 13-2 Three-point bending beam specimens

Specimen	Aggregate Category	Aggregate size d_{max} (mm)	Design Strength	$S \times D \times B$ (mm) Specimen size	Initial crack length a_0 (mm)	Cube compressive strength f_{cu} (MPa)	Axial compressive strength f_c (MPa)	Splitting tensile strength f_{ts} (MPa)	Elastic modulus E (GPa)	Poisson's ratio v
SL 1	Small gravel	10	C20	1200×300×120	120	43.8	30.3	4.12	26.7	0.171
SL 1y				1600×400×120	160					
SL 2				400×100×120	40					
SL 3				600×150×120	60					
SL 4				800×200×120	80					
SL 5	Small gravel One graded	20	C20	1200×300×120	120	43.3	32.1	2.99	33.2	0.208
SL 6				1600×400×120	160					
SL 7				2000×500×120	200					
SL 8			C30	1200×300×120	120	53.7	38.6	3.53	31.1	0.177
SL 9			C40			56.4	43.3	5.55	30.5	0.184
SL 10			C50			50.2	35.4	3.95	34.8	0.184

13.3 Determination of Initial Cracking Load P_{ini}

continue

Specimen	Aggregate Category	Aggregate size d_{max} (mm)	Design Strength	$S \times D \times B$ (mm) Specimen size	Initial crack length a_0 (mm)	Cube compressive strength f_{cu} (MPa)	Elastic modulus E (GPa)	Splitting tensile strength f_{ts} (MPa)	Axial compressive strength f_c (MPa)	Poisson's ratio v
SL 11	Two graded	40	C20	1200×300×120	120	51.2	34.6	3.39	33.3	0.220
SL 43				1600×400×240	160	56.2	35	3.14	37.1	0.196
SL 44	Large aggregate Fully graded	80	C25	1800×450×240	180	47.1	32.3	2.70	41.0	0.222
SL 45				2000×500×240	200	47.1	32.3	2.70	41.0	0.222
SL 46				2200×550×240	220	56.2	35	3.14	37.1	0.196
SL 47	Wet-screening	40	C25	800×200×120	80	56.2	35	3.14	37.1	0.196
SL 48				1000×250×120	100	56.2	35	3.14	37.1	0.196
SL 49				1200×300×120	120	47.1	32.3	2.70	41.0	0.222
SL 50				1600×400×120	160	47.1	32.3	2.70	41.0	0.222

Table 13-3 Wedge splitting specimens

Specimen	Aggregate Category	Aggregate size d_{max} (mm)	Design Strength	$S \times D \times B$ (mm) Specimen size	Initial crack length a_0 (mm)	Cube compressive strength f_{cu} (MPa)	Elastic modulus E (GPa)	Splitting tensile strength f_{ts} (MPa)	Axial compressive strength f_c (MPa)	Poisson's ratio v
WS 12	Small gravel	10	C20	300×300×200	150	43.8	26.7	4.12	30.3	0.171
WS 12y				600×600×200	300					
WS 13				300×300×200	150					
WS 14				600×600×200	300					
WS 15	Small gravel One graded	20	C20	800×800×200	400	43.4	33.4	2.76	34.2	0.202
WS 16				1000×1000×200	500					
WS 17				1200×1200×200	600					
WS 18			C30	300×300×200	150	48.1	29.6	3.81	38.6	0.190
WS 19			C40			56.4	30.5	5.55	43.3	0.184
WS 20			C50			50.2	34.8	3.95	35.4	0.184

continue

Specimen	Aggregate Category	Aggregate size d_{max} (mm)	Design Strength	$S \times D \times B$ (mm) Specimen size	Initial crack length a_0 (mm)	Cube compressive strength f_{cu} (MPa)	Elastic modulus E (GPa)	Splitting tensile strength f_{ts} (MPa)	Axial compressive strength f_c (MPa)	Poisson's ratio v
WS 21	Two graded	40	C20	300×300×200	150	50.4	31.5	2.98	31.5	0.190
WS 22				450×450×250	225	/	/	/	/	/
WS 23	Large aggregate Fully graded	80	C20	600×600×250	300	40.8	29.1	3.04	34.3	0.189
WS 24				800×800×250	400	40.8	29.1	3.04	34.3	0.189
WS 25				1000×1000×250	500	40.8	29.1	3.04	34.3	0.189
WS 26				1200×1200×250	600	/	/	/	/	/
WS 32	Wet-screening	40	20	300×300×200	150	40.8	29.1	3.04	34.3	0.189
WS 33				600×600×200	300	40.8	29.1	3.04	34.3	0.189
WS 34				800×800×200	400	/	/	/	/	/
WS 35				1000×1000×200	500	/	/	/	/	/

* Note: Label"/"for no conducting basic mechanics performance test.

13.3 Determination of Initial Cracking Load P_{ini}

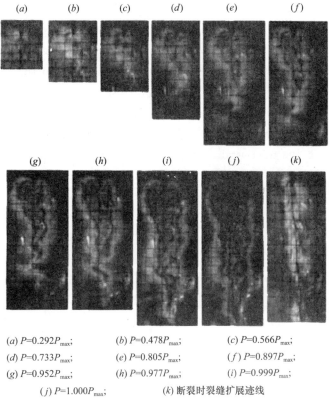

(a) $P=0.292P_{max}$;
(b) $P=0.478P_{max}$;
(c) $P=0.566P_{max}$;
(d) $P=0.733P_{max}$;
(e) $P=0.805P_{max}$;
(f) $P=0.897P_{max}$;
(g) $P=0.952P_{max}$;
(h) $P=0.977P_{max}$;
(i) $P=0.999P_{max}$;
(j) $P=1.000P_{max}$;
(k) 断裂时裂缝扩展迹线

Fig. 13-1 Process of crack propagation and the development of FPZ by photoelastic coating

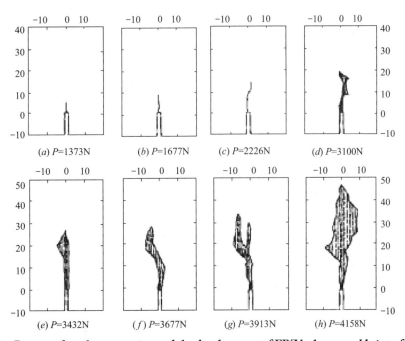

Fig. 13-2 Process of crack propagation and the development of FPZ by laser speckle interferometry

As shown in Fig. 13-3, it can be seen the strain of strain gauge at the tip of pre-crack increased with increment of the load and tended to linear relationship at the initial segment of load-strain curve. When the load continued to increase resulting in concrete cracking, the concrete deformation energy was released on the both sides of the crack and the strain value is retracted. The figures show the turning point appeared after the strain of initial strain gauge reaches a maximum value, then the strain value begins to decreased. Therefore, the load corresponding to the peak strain is the initial cracking load of the ligament. When the load continues to increase, the next strain gauge arranged along the ligament appears the same phenomenon.

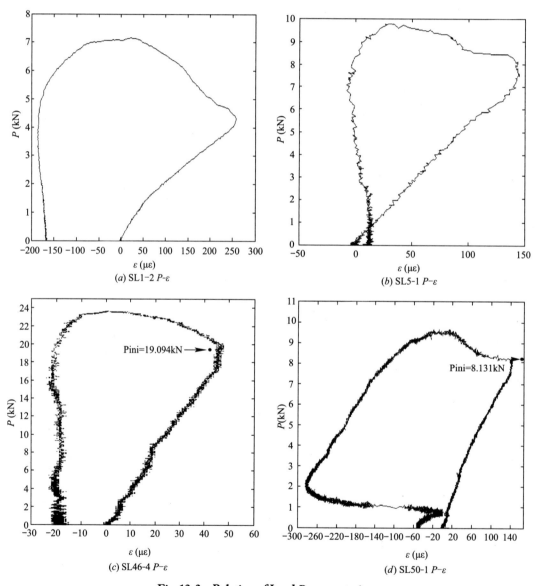

Fig. 13-3 Relation of Load P versus strain ε

13.4 Calculation Procedures and Results of the Double-K Fracture Parameters

13.4.1 Calculation procedures

By the tested *P-CMOD* curve, the geometric parameters of the specimen, and the cracking load P_{ini} obtained from the above-mentioned, the initiation toughness $K_{\text{IC}}^{\text{ini}}$ and unstable fracture toughness $K_{\text{IC}}^{\text{un}}$ can be calculated according to the following steps:

13.4.2 Calculation of TPB specimen

(a) The initial flexibility C_i is measured from the rising section of the *P-CMOD* curve. The elastic modulus E of concrete can be calculated by the equation (13-1). Where H_0 is the knife-edge thickness for the clip type extensometer.

$$E = \frac{1}{BC_i}\left[3.70 + 32.60 \tan^2\left(\frac{\pi}{2}\frac{a_0 + H_0}{D + H_0}\right)\right] \tag{13-1}$$

(b) The critical crack length a_c can be calculated by plugging the maximum load $P_{\max}$ of the *P-CMOD* curve, critical crack mouth opening displacement $CMOD_c$ and the elastic modulus E obtained from formula (1) into equation (13-2).

$$a_c = \frac{2}{\pi}(D + H_0)\arctan\sqrt{\frac{B \cdot E \cdot CMOD_c}{32.6 P_{\max}} - 0.1135} - H_0 \tag{13-2}$$

(c) There are two kinds of stress intensity factors equation (13-3) and (13-5) for a TPB specimen. Tada, Shah and the specification Metallic Materials-Determination of Plane-strain Fracture Toughness GB 4161-84 [18] adopted equation (13-3), and the ASTM E399-72 adopted equation (13-5).

$$K_{\text{IC}} = \frac{1.5(P + W/2)S}{BD^2}\sqrt{a}F_1(V) \tag{13-3}$$

$$F_1(V) = \frac{1.99 - V(1-V)(2.15 - 3.93V + 2.7V^2)}{(1+2V)(1-V)^{3/2}} \tag{13-4}$$

$$K_I = \frac{(P + W/2)S}{BD^{3/2}}F_1(V) \tag{13-5}$$

$$F_1(V) = 2.9V^{1/2} - 4.6V^{3/2} + 21.8V^{5/2} - 37.6V^{7/2} + 38.7V^{9/2} \tag{13-6}$$

When the $K_{\text{IC}}^{\text{um}}$ is calculated, let $P = P_{\max}$ and $V = a_c/D$. When the $K_{\text{IC}}^{\text{ini}}$ is calculated, let $P = P_{\text{ini}}$ and $V = a_0/D$.

13.4.3 Calculation of WS specimen

(a) The initial flexibility C_i is measured from rising section of P-$CMOD$ curve. The elastic modulus E can be obtained by the equation (13-7).

$$E = \frac{1}{BC_i}\left[13.18\left(1 - \frac{a_0 + H_0}{D + H_0}\right)^2 - 9.16\right] \quad (13\text{-}7)$$

(b) The critical crack length a_c can be calculated by the equation (13-8).

$$a_c = (D + H_0)\left(1 - \sqrt{\frac{13.18}{B \cdot E \cdot CMOD_c / P_{max} + 9.16}}\right) - H_0 \quad (13\text{-}8)$$

(c) The equation (13-9) can be adopted to calculate the stress intensity factor of WS specimen.

$$K_{IC} = \frac{3.675P[1 - 0.12(V - 0.45)]}{B\sqrt{D}(1 - V)^{3/2}} \quad (13\text{-}9)$$

When the K_{IC}^{um} is calculated, let $P = P_{max}$ and $V = a_c/D$. When the K_{IC}^{ini} is calculated, let $P = P_{ini}$ and $V = a_0/D$.

13.4.4 Experimental results

In accordance with the above-mentioned methods and steps, the calculated results are shown in Table 13-4 and 13-5. Fig. 13-4(a)-(c) are the curves of fracture toughness via specimen height of one graded aggregate concrete TPB specimens (six series SL2-SL7), fully graded large aggregate concrete TPB specimens (four series SL43-SL46) and wet-screening concrete TPB specimens (four series SL47-SL50). Fig. 13-5(a)-(c) are the curves of fracture toughness via specimen height of one graded aggregate concrete WS specimens (five series WS13-WS17), fully graded large aggregate concrete WS specimens (five series WS22-WS26) and wet-screening concrete WS specimens (four series WS32-WS35).

13.5 Discussion on Wedge-splitting Test Results

Wedge-splitting method is an improved method based on advantages of compact tension method and Wedge-Opening-Loading (WOL) type compact tension specimen. The WS specimen has the following advantages: eliminating the effect of specimen dead-weight; the smallest volume specimen to have the same measurement capability. The structural concrete fracture parameters can be measured by core sampling in existing structures.

The following analyzed the effect of loading mode on crack tip stress field and calculation of fracture parameters. To eliminate the influence of a specimen dead-weight, the bearing of the WS

13.5 Discussion on Wedge-splitting Test Results

Table 13-4 Double-K fracture parameters of TPB specimens

Specimen	B (mm)	D (mm)	S (mm)	a_0 (mm)	$CMOD_c$ (mm)	P_{max} (kN)	P_{ini} (kN)	P_{max}/P_{ini}	C_i (m/N)	E (GPa) (1)	E (GPa) Tested value	a_c (mm) (2)	γ_c	K_{IC}^{un} (MPa·m$^{1/2}$) (3)	K_{IC}^{ini} (MPa·m$^{1/2}$) (3)	K_{IC}^{un} (MPa·m$^{1/2}$) (5)	K_{IC}^{ini} (MPa·m$^{1/2}$) (5)
\multicolumn{18}{c}{Small size aggregate specimens d_{max}=10mm}
SL 1-2	120	300	1200	120	0.09118	7.176	4.2	0.585	0.00745	23.39	26.7	149.05315	0.4968	1.2293	0.5653	1.3084	0.6238
SL 1-3	120	300	1200	120	0.13322	7.401	5.3975	0.729	0.007352	23.7	26.7	167.85845	0.5595	1.5596	0.7097	1.6614	0.7681
SL 1-4	120	300	1200	120	0.09914	7.031	3.59	0.511	0.007254	24.02	26.7	155.96597	0.5199	1.2989	0.4917	1.3855	0.5502
Average	120	300	1200	120	0.1078	7.2027	4.3958	0.6083	0.0074	23.7	26.7	157.6259	0.5254	1.3626	0.5889	1.4518	0.6474
SL 1y-1	120	400	1600	160	0.11569	11.008	7.194	0.654	0.00555	31.39	26.7	206.12303	0.5153	1.7502	0.8418	1.8807	0.9319
SL 1y-2	120	400	1600	160	0.09357	9.004	7	0.777	0.0058822	29.62	26.7	201.23471	0.5031	1.3986	0.8216	1.5229	0.9116
SL 1y-3	120	400	1600	160	0.11897	10.863	8.2687	0.761	0.0056644	30.76	26.7	207.58045	0.519	1.7495	0.9541	1.8819	1.0441
Average	120	400	1600	160	0.1094	10.2917	7.4876	0.7307	0.0057	30.59	26.7	204.9794	0.5124	1.6328	0.8725	1.7618	0.9625
\multicolumn{18}{c}{One graded aggregate specimens}
SL 2-1	120	100	400	40	0.05	6	4.999	0.833		27.76	33.2	45.181774	0.4518	1.4663	1.0556	1.4787	1.0665
SL 2-2	120	100	400	40	0.05	4.5	4	0.889		27.76	33.2	50.342373	0.5034	1.2921	0.8469	1.3088	0.8579
SL 2-3	120	100	400	40	0.04461	3.475	3.119	0.898	0.0068625	25.39	33.2	51.316612	0.5132	1.0332	0.6629	1.0506	0.6739
SL 2-4	120	100	400	40	0.05637	4.60645	3.7929	0.823	0.005784	30.12	33.2	53.460669	0.5346	1.4646	0.8036	1.4853	0.8146
Average	120	100	400	40	0.0502	4.6454	3.9777	0.8608		27.76	33.2	50.0754	0.5008	1.3141	0.8423	1.3309	0.8532
SL 3-1	120	150	600	60	0.06506	7.318	6.82	0.932	0.0056367	30.91	33.2	72.435053	0.4829	1.616	1.184	1.6428	1.2043
SL 3-2	120	150	600	60	0.068	5.774	5.1586	0.893	0.005392	32.31	33.2	81.01367	0.5401	1.5411	0.9007	1.5772	0.9211
SL 3-3	120	150	600	60	0.06568	6.176	5.662	0.917	0.0056033	31.1	33.2	77.355784	0.5157	1.5179	0.9865	1.5497	1.0069
SL 3-4	120	150	600	60	0.07374	6.215	5.69	0.916	0.0052695	33.07	33.2	81.795724	0.5453	1.6862	0.9913	1.7236	1.0117
Average	120	150	600	60	0.0681	6.3708	5.8327	0.9145	0.0055	31.85	33.2	78.1501	0.521	1.5903	1.0156	1.6233	1.036

13 Experimental Research on Double-K Fracture Parameters of Dam Concrete with Various Aggregate Gradation

continue

Specimen	B (mm)	D (mm)	S (mm)	a_0 (mm)	$CMOD_c$ (mm)	P_{max} (kN)	P_{ini} (kN)	P_{max}/P_{ini}	C_i (m/N)	E (GPa) (1)	E (GPa) Tested value	a_c (mm) (2)	γ_c	K_{IC}^{un} (MPa·m$^{1/2}$) (3)	K_{IC}^{ini} (MPa·m$^{1/2}$) (3)	K_{IC}^{un} (MPa·m$^{1/2}$) (5)	K_{IC}^{ini} (MPa·m$^{1/2}$) (5)
SL 4-1	120	200	800	80	0.08235	8.573	7.433	0.867	0.0057833	30.13	33.2	98.426317	0.4921	1.7015	1.1299	1.7444	1.1615
SL 4-3	120	200	800	80	0.04957	7.3481	6.216	0.846	0.0037815	46.08	33.2	100.96076	0.5048	1.5243	0.9502	1.5697	0.9818
SL 4-4	120	200	800	80	0.07206	7	5.925	0.846	0.0053922	32.31	33.2	103.3469	0.5167	1.5113	0.9072	1.5592	0.9388
Average	120	200	800	80	0.068	7.6404	6.5247	0.853	0.005	36.17	33.2	100.9113	0.5046	1.579	0.9958	1.6244	1.0274
SL 5-1	120	300	1200	120	0.06863	9.83113	7.556	0.769	0.0039916	43.65	33.2	150.389	0.5013	1.6788	0.97	1.7597	1.0284
SL 5-3	120	300	1200	120	0.09412	10.31169	8.345	0.809	0.005112	34.08	33.2	151.48786	0.505	1.7776	1.0652	1.8598	1.1235
SL 5-4	120	300	1200	120	0.09627	8.907	6.518	0.732	0.0053921	32.31	33.2	157.57359	0.5252	1.6519	0.8449	1.7411	0.9032
Average	120	300	1200	120	0.0863	9.6833	7.473	0.77	0.0048	36.68	33.2	153.1502	0.5105	1.7028	0.96	1.7869	1.0184
SL 6-1	120	400	1600	160	0.07647	12.9	7.764	0.602	0.0038726	44.99	33.2	191.01457	0.4775	1.801	0.9014	1.9144	0.9914
SL 6-3	120	400	1600	160	0.10196	14.2069	9.384	0.661	0.004657	37.41	33.2	191.4972	0.4787	1.9793	1.0706	2.0932	1.1605
SL 6-4	120	400	1600	160	0.087745	9.3553	6.144	0.657	0.0055143	31.6	33.2	198.54362	0.4964	1.4178	0.7322	1.539	0.8222
Average	120	400	1600	160	0.0887	12.1541	7.764	0.64	0.0047	38	33.2	193.6851	0.4842	1.7327	0.9014	1.8489	0.9914
SL 7-1	120	500	2000	200	0.1345	14.4828	9.366	0.647	0.0047609	36.6	33.2	260.25032	0.5205	2.1235	1.0014	2.309	1.1273
SL 7-2	120	500	2000	200	0.1273	9.003	5.65	0.628	0.0053921	32.31	33.2	285.51268	0.571	1.6525	0.6542	1.8748	0.7802
SL 7-3	120	500	2000	200	0.11657	14.204	11.19	0.788	0.0054288	32.1	33.2	237.64577	0.4753	1.808	1.1718	1.9652	1.2976
SL 7-4	120	500	2000	200	0.11373	16.4432	12.978	0.789	0.0045207	38.54	33.2	238.72313	0.4774	2.0819	1.3388	2.2403	1.4646
Average	120	500	2000	200	0.123	13.5333	9.796	0.713	0.005	34.89	33.2	255.533	0.5111	1.9165	1.0416	2.0973	1.1674
SL 8-1	120	300	1200	120	0.1074	12.206	9.389	0.769	0.00522	33.38	31.1	148.42711	0.4948	2.0229	1.1911	2.1019	1.2494
SL 8-2	120	300	1200	120	0.08931	9.8526	8.232	0.836	0.0049838	34.96	31.1	152.46682	0.5082	1.7199	1.0516	1.8032	1.1099

13.5 Discussion on Wedge-splitting Test Results

continue

Specimen	B (mm)	D (mm)	S (mm)	a_0 (mm)	$CMOD_c$ (mm)	P_{max} (kN)	P_{ini} (kN)	P_{max}/P_{ini}	C_i (m/N)	E (GPa) (1)	E (GPa) Tested value	a_c (mm) (2)	γ_c	K_{IC}^{un} (MPa·m$^{1/2}$) (3)	K_{IC}^{ini} (MPa·m$^{1/2}$) (3)	K_{IC}^{un} (MPa·m$^{1/2}$) (5)	K_{IC}^{ini} (MPa·m$^{1/2}$) (5)
SL 8-3	120	300	1200	120	0.08422	9.895	7.8	0.788	0.0041962	41.52	31.1	158.19139	0.5273	1.8382	0.9995	1.9285	1.0578
SL 8-4	120	300	1200	120	0.1287	9.976	6.3676	0.638	0.0060379	28.86	31.1	160.89264	0.5363	1.9093	0.8267	2.0032	0.8851
Average	120	300	1200	120	0.1024	10.4824	7.9472	0.7578	0.0051	34.68	31.1	154.9945	0.5166	1.8726	1.0172	1.9592	1.0756
SL 9-2	120	300	1200	120	0.08616	11.002		0	0.0054463	31.99	30.5	139.88222	0.4663	1.6773	0.0587	1.748	0.1174
SL 9-3	120	300	1200	120	0.1	10.747	8.9697	0.835	0.0056911	30.62	30.5	146.79763	0.4893	1.7602	1.1406	1.8373	1.1988
SL 9-4	120	300	1200	120	0.07684	10.998	8.6522	0.787	0.0044398	39.24	30.5	144.74749	0.4825	1.7617	1.1023	1.8368	1.1605
Average	120	300	1200	120	0.0877	10.9157	5.874	0.5407	0.0052	33.95	30.5	143.8091	0.4794	1.7331	0.7672	1.8074	0.8256
SL 10-1	120	300	1200	120	0.08854	10.513	9.1077	0.866	0.0051131	34.08	34.8	147.19336	0.4906	1.7306	1.1572	1.8082	1.2154
SL 10-2	120	300	1200	120	0.11732	12.298	10.665	0.867	0.0048434	35.97	34.8	156.65698	0.5222	2.2257	1.345	2.3148	1.4032
SL 10-3	120	300	1200	120	0.096665	11.247		0	0.0053782	32.4	34.8	145.57437	0.4852	1.8153	0.0587	1.8913	0.1174
SL 10-4	120	300	1200	120	0.11897	9.8	8.32	0.849	0.00526	33.12	34.8	164.85659	0.5495	1.964	1.0622	2.0632	1.1205
Average	120	300	1200	120	0.1054	10.9645	7.0232	0.6455	0.0051	33.89	34.8	153.5703	0.5119	1.9339	0.9058	2.0194	0.9641
Two graded aggregate specimens																	
SL 11-2	120	300	1200	120	0.10534	9.5889	6.49	0.677	0.0043556	40	34.6	169.35719	0.5645	2.0279	0.8415	2.1331	0.8998
SL 11-3	120	300	1200	155.1	0.13046	6.292	4.221	0.671	0.0096217	34.6	34.6	192.76213	0.6425	1.8488	0.8054	1.9731	0.8902
SL 11-4	120	300	1200	120	0.08489	9.408	7.073	0.752	0.0052117	33.43	34.6	149.85186	0.4995	1.6008	0.9118	1.6811	0.9701
Average	120	300	1200	131.7	0.1069	8.4296	5.928	0.7	0.0064	36.01	34.6	170.6571	0.5689	1.8258	0.8529	1.9291	0.92
Fully graded large aggregate specimens																	
SL 43-1	240	400	1600	160	0.1766	17.62	11.826	0.671	0.0030985	28.12	35	242.08877	0.6052	1.9626	0.708	2.1429	0.7981

13 Experimental Research on Double-K Fracture Parameters of Dam Concrete with Various Aggregate Gradation

continue

Specimen	B (mm)	D (mm)	S (mm)	a_0 (mm)	$CMOD_c$ (mm)	P_{max} (kN)	P_{ini} (kN)	P_{max}/P_{ini}	C_i (m/N)	E (GPa) (1)	E (GPa) Tested value	a_c (mm) (2)	γ_c	K_{IC}^{un} (MPa·m$^{1/2}$) (3)	K_{IC}^{ini} (MPa·m$^{1/2}$) (3)	K_{IC}^{un} (MPa·m$^{1/2}$) (5)	K_{IC}^{ini} (MPa·m$^{1/2}$) (5)
SL 43-2	240	400	1600	160	0.12331	17.656	10.3	0.583	0.0027974	31.14	35	225.12449	0.5628	1.6793	0.6283	1.8354	0.7185
SL 43-3	240	400	1600	160	0.13137	23.414	15.692	0.67	0.0024104	36.14	35	220.38889	0.551	2.0893	0.9099	2.24	0.9999
SL 43-4	240	400	1600	160	0.08117	13.629	11.06	0.64	0.0025892	33.65	35	219.58065	0.549	1.2673	0.668	1.4144	0.7581
Average	240	400	1600	160	0.1281	18.0798	12.2195	0.641	0.0027	32.26	35	226.7957	0.567	1.7496	0.7286	1.9082	0.8187
SL 44-1	240	450	1800	180	0.15367	20.909	15.2345	0.729	0.0026559	32.8	32.3	260.9708	0.5799	2.0071	0.858	2.205	0.9655
SL 44-2	240	450	1800	180	0.13322	21.864	16.4	0.75	0.0027956	31.16	32.3	242.87646	0.5397	1.8129	0.9154	1.9833	1.0229
SL 44-3	240	450	1800	180	0.10348	24.587	15.239	0.62	0.0024299	35.85	32.3	224.81254	0.4996	1.769	0.8582	1.9155	0.9657
SL 44-4	240	450	1800	180	0.13508	21.824	16.336	0.749	0.0025177	34.6	32.3	252.09788	0.5602	1.9432	0.9122	2.1276	1.0197
Average	240	450	1800	180	0.1314	22.296	15.8024	0.712	0.0026	33.6	32.3	245.1894	0.5449	1.8831	0.886	2.0579	0.9935
SL 45-1	240	500	2000	200	0.16483	25.466	9.22078	0.362	0.0029101	29.94	32.3	271.60199	0.5432	2.0382	0.5571	2.2402	0.6831
SL 45-2	240	500	2000	200	0.19643	20.993	16.528	0.787	0.0029335	29.7	32.3	301.49332	0.603	2.1311	0.8984	2.3801	1.0244
SL 45-3	240	500	2000	200	0.14933	26.631	20.64935	0.775	0.0027198	32.03	32.3	265.06802	0.5301	2.0308	1.0909	2.223	1.2168
SL 45-4	240	500	2000	200	0.16297	21.72	14.0356	0.646	0.0026517	32.85	32.3	291.79846	0.5836	2.0407	0.782	2.2744	0.908
Average	240	500	2000	200	0.1684	23.7025	15.1084	0.6425	0.0028	31.13	32.3	282.4904	0.565	2.0602	0.8321	2.2794	0.9581
SL 46-1	240	550	2200	220	0.1673	23.165	18.8486	0.814	0.0028571	30.49	35	310.69034	0.5649	1.9675	0.9852	2.2186	1.1306
SL 46-2	240	550	2200	220	0.18651	27.149	14	0.516	0.0028702	30.35	35	305.63626	0.5557	2.1915	0.7693	2.4351	0.9147
SL 46-4	240	550	2200	220	0.11835	23.792	19.094	0.803	0.0022106	39.41	35	299.8597	0.5452	1.8799	0.9962	2.1133	1.1415
Average	240	550	2200	220	0.1574	24.702	17.3142	0.711	0.0026	33.42	35	305.3954	0.5553	2.013	0.9169	2.2557	1.0623

13.5 Discussion on Wedge-splitting Test Results

continue

Specimen	B (mm)	D (mm)	S (mm)	a_0 (mm)	$CMOD_c$ (mm)	P_{max} (kN)	P_{ini} (kN)	P_{max}/P_{ini}	C_i (m/N)	E (GPa) (1)	E (GPa) Tested value	a_c (mm) (2)	γ_c	K_{IC}^{un} (MPa·m$^{1/2}$) (3)	K_{IC}^{ini} (MPa·m$^{1/2}$) (3)	K_{IC}^{un} (MPa·m$^{1/2}$) (5)	K_{IC}^{ini} (MPa·m$^{1/2}$) (5)
\multicolumn{18}{l}{Wet-screening specimens}
SL 47-1	120	200	800	80	0.08179	6.919	6.312	0.912	0.005375	32.42	35	108.25517	0.5413	1.6218	0.9643	1.6756	0.9959
SL 47-2	120	200	800	80	0.08427	7.718	5.453	0.707	0.0052833	32.98	35	106.10944	0.5305	1.7393	0.8375	1.7908	0.8691
SL 47-3	120	200	800	80	0.08173	7.537	6.212	0.824	0.0053072	32.83	35	105.71475	0.5286	1.6884	0.9496	1.7394	0.9812
SL 47-4	120	200	800	80	0.08613	6.818	5.303	0.778	0.0059795	29.14	35	106.87197	0.5344	1.5618	0.8153	1.6138	0.8469
Average	120	200	800	80	0.0835	7.248	5.82	0.8053	0.0055	31.84	35	106.7378	0.5337	1.6528	0.8917	1.7049	0.9233
SL 48-1	120	250	1000	100	0.07469	8.525	5.7896	0.679	0.0049403	35.27	35	125.9406	0.5038	1.5922	0.8096	1.6547	0.8539
SL 48-2	120	250	1000	100	0.09047	10.372	7.304	0.704	0.005137	33.92	35	124.01379	0.4961	1.8775	1.0097	1.9382	1.0539
SL 48-3	120	250	1000	100	0.07188	7.3287	5.441	0.742	0.0045118	38.62	35	134.81954	0.5393	1.5481	0.7635	1.6206	0.8079
SL 48-4	120	250	1000	100	0.081174	9.0589	7.762	0.857	0.0051714	33.69	35	124.91259	0.4997	1.6661	1.0702	1.7276	1.1144
Average	120	250	1000	100	0.0796	8.8212	6.5742	0.7455	0.0049	35.37	35	127.4216	0.5097	1.671	0.9133	1.7353	0.9575
SL 49-1	120	300	1200	120	0.11278	9.9634	7.826085	0.785	0.0051615	33.76	32.3	162.23741	0.5408	1.9362	1.0026	2.0319	1.0609
SL 49-2	120	300	1200	120	0.081	10.66	7.69565	0.722	0.0047701	36.53	32.3	145.40212	0.4847	1.7215	0.9869	1.7972	1.0452
SL 49-3	120	300	1200	120	0.07684	9.539	6.256	0.656	0.0050041	34.82	32.3	145.96406	0.4865	1.5574	0.8133	1.6336	0.8716
Average	120	300	1200	120	0.0902	10.0541	7.2592	0.721	0.005	35.03	32.3	151.2012	0.504	1.7384	0.9343	1.8209	0.9926
SL 50-1	120	400	1600	160	0.08317	9.479	8.131	0.858	0.0046478	37.49	32.3	205.90769	0.5148	1.5221	0.9397	1.6521	1.0297
SL 50-2	120	400	1600	160	0.087	9.109	6.9441	0.762	0.0051736	33.68	32.3	204.33691	0.5108	1.4491	0.8157	1.5771	0.9058
SL 50-3	120	400	1600	160	0.07387	11.433	8.963	0.784	0.0044031	39.57	32.3	187.97366	0.4699	1.5723	1.0266	1.6827	1.1166
Average	120	400	1600	160	0.0813	10.007	8.0127	0.8013	0.0047	36.91	32.3	199.4061	0.4985	1.5145	0.9273	1.6373	1.0174

13 Experimental Research on Double-K Fracture Parameters of Dam Concrete with Various Aggregate Gradation

Double-K fracture parameters of WS specimens   Table 13-5

Specimen	B (mm)	D (mm)	a_0 (mm)	$CMOD_c$ (mm)	P_{max} (kN)	P_{ini} (kN)	P_{max}/P_{ini}	C_i (m/N)	E (GPa) (7)	E (GPa) Tested value	a_c (mm) (8)	γ_c	K_{IC}^{un} (MPa·m$^{1/2}$) (9)	K_{IC}^{ini} (MPa·m$^{1/2}$) (9)
\multicolumn{15}{l}{Small size aggregate specimens d_{max}=10mm}
WS 12-1	200	300	150	0.181368	10.311	7.96	0.77	0.009679	25.295	26.7	184.6	0.6152	1.4205	0.7508
WS 12-3	200	300	150	0.15114	8.095	4.5543	0.56	0.009121	26.842	26.7	190.7	0.6355	1.2068	0.4296
WS 12-4	200	300	150	0.17686	8.562	5.8942	0.69	0.008915	27.461	26.7	196.7	0.6557	1.387	0.5559
Average	200	300	150	0.169789	8.989333	6.136167	0.67	0.009238	26.53	26.7	190.7	0.6355	1.3381	0.5788
WS 12y-1	200	600	300	0.2607	17.5325	13.27413	0.76	0.007663	30.162	26.7	375.4	0.6257	1.7782	0.8853
WS 12y-2	200	600	300	0.21636	15.899	11.284	0.71	0.008763	26.376	26.7	351.8	0.5864	1.3947	0.7526
WS 12y-4	200	600	300	0.34061	18	14.032	0.78	0.008486	27.24	26.7	389.3	0.6488	2.0024	0.9358
Average	200	600	300	0.272557	17.14383	12.86338	0.75	0.008304	27.93	26.7	372.2	0.6203	1.7251	0.8579
\multicolumn{15}{l}{One graded aggregate specimens}
WS 13-1	200	300	150	0.15113	12.173	7.181	0.59	0.00675	36.269	33.4	185.2	0.6173	1.6904	0.6773
WS 13-2	200	300	150	0.17543	12.80104	10.9158	0.85	0.006443	37.998	33.4	192.6	0.6419	1.9578	1.0296
WS 13-4	200	300	150	0.14611	11.49158	7.9091	0.69	0.005972	40.993	33.4	192.6	0.642	1.7586	0.746
Average	200	300	150	0.157557	12.15521	8.668633	0.71	0.006388	38.42	33.4	190.1	0.6337	1.8023	0.8176
WS 14-1	200	600	300	0.2402	25.55	19.3083	0.76	0.006127	37.724	33.4	350.5	0.5842	2.2241	1.2877
WS 14-2	200	600	300	0.25282	22.667	18.484	0.82	0.005185	44.578	33.4	385.8	0.6429	2.4616	1.2328
WS 14-4	200	600	300	0.2666	23.4081	18	0.77	0.007617	30.345	33.4	347.6	0.5794	2.0042	1.2005
Average	200	600	300	0.253207	23.87503	18.59743	0.78	0.00631	37.55	33.4	361.3	0.6022	2.23	1.2403
WS 15-1	200	800	400	0.3263	30.7575	23.788	0.77	0.007549	30.173	33.4	454.1	0.5677	2.1914	1.374
WS 15-2	200	800	400	0.26818	31.13636	24.5455	0.79	0.006514	34.967	33.4	444.8	0.556	2.1348	1.4177
WS 15-3	200	800	400	0.3006	29.3506	17.4025	0.59	0.007228	31.513	33.4	455.4	0.5692	2.1021	1.0051

13.5 Discussion on Wedge-splitting Test Results

continue

Specimen	B (mm)	D (mm)	a_0 (mm)	$CMOD_c$ (mm)	P_{max} (kN)	P_{ini} (kN)	P_{max}/P_{ini}	C_i (m/N)	E (GPa) (7)	E (GPa) Tested value	a_c (mm) (8)	γ_c	K_{IC}^{un} (MPa·m$^{1/2}$) (9)	K_{IC}^{ini} (MPa·m$^{1/2}$) (9)
Average	200	800	400	0.29836	30.41482	21.912	0.72	0.007097	32.22	33.4	451.4	0.5643	2.1428	1.2656
WS 16-1	200	1000	500	0.39658	42.137	32.495	0.77	0.005706	39.568	33.4	597.1	0.5971	2.9739	1.6787
WS 16-2	200	1000	500	0.38108	39	30.478	0.78	0.005613	40.226	33.4	606.7	0.6067	2.8505	1.5745
WS 16-3	200	1000	500	0.3079	31.494	24.235	0.77	0.005817	38.812	33.4	600.4	0.6004	2.2499	1.252
WS 16-4	200	1000	500	0.39534	35.84	11.075	0.31	0.006725	33.571	33.4	596.1	0.5961	2.5204	0.5721
Average	200	1000	500	0.370225	37.11775	23.58085	0.63	0.005965	38.04		600.1	0.6001	2.6487	1.2182
WS 17-2	200	1200	600	0.51462	46.326	33.368	0.72	0.00596	37.656	33.4	742.2	0.6185	3.2307	1.5736
WS 17-3	200	1200	600	0.43933	55.183	36.699	0.67	0.006212	36.129	33.4	659.9	0.5499	3.0284	1.7307
WS 17-4	200	1200	600	0.4428	50.3551	40.045	0.8	0.005686	39.469	33.4	702.4	0.5853	3.1117	1.8885
Average	200	1200	600	0.465583	50.62137	36.704	0.73	0.005953	37.75	33.4	701.5	0.5846	3.1236	1.7309
WS 18-1	200	300	150	0.1634	13.6774	12.085	0.88	0.008825	27.742	29.6	168.3	0.5611	1.5572	1.1398
WS 18-2	200	300	150	0.1635	13.8447	13.2301	0.96	0.006618	36.996	29.6	183.6	0.612	1.8847	1.2478
WS 18-3	200	300	150	0.2078	14.4483	11.569	0.8	0.009133	26.805	29.6	176.9	0.5896	1.8126	1.0912
WS 18-4	200	300	150	0.1791	14.3103	13.7586	0.96	0.007114	34.414	29.6	182.9	0.6095	1.9298	1.2977
Average	200	300	150	0.17845	14.07018	12.66068	0.9	0.007923	31.49	29.6	177.9	0.5931	1.7961	1.1941
WS 19-1	200	300	150	0.1799	10.7069	8.1964	0.77	0.00755	32.426	30.5	194.8	0.6493	1.688	0.7731
WS 19-2	200	300	150	0.1632	9.643	7.815	0.81	0.008067	30.349	30.5	191.9	0.6396	1.4613	0.7371
WS 19-3	200	300	178.2	0.163423	9.8696	7.718	0.78	0.012954	30.5	30.5	191	0.6368	1.4789	0.9838
Average	200	300	150	0.168841	10.07317	7.9098	0.79	0.009524	31.09	30.5	192.6	0.6419	1.5427	0.8313
WS 20-1	200	300	150	0.15395	15.0769	13.15	0.87	0.006179	39.621	34.8	179.5	0.5983	1.9517	1.2403
WS 20-2	200	300	150	0.18961	14.8907	11.568	0.78	0.007647	32.015	34.8	179.9	0.5997	1.9372	1.0911
WS 20-4	200	300	150	0.1703	14.6923	11.749	0.8	0.008158	30.01	34.8	171.1	0.5704	1.7251	1.1082
Average	200	300	150	0.171287	14.88663	12.15567	0.82	0.007328	33.88	34.8	176.9	0.5895	1.8713	1.1465

continue

Specimen	B (mm)	D (mm)	a_0 (mm)	$CMOD_c$ (mm)	P_{max} (kN)	P_{ini} (kN)	P_{max}/P_{ini}	C_i (m/N)	E (GPa) (7)	E (GPa) Tested value	a_c (mm) (8)	γ_c	K_{IC}^{un} (MPa·m$^{1/2}$) (9)	K_{IC}^{ini} (MPa·m$^{1/2}$) (9)
					Two graded aggregate specimens									
WS 21-1	200	300	150	0.15	12.6098	11.69	0.93	0.006503	37.646	31.5	184.9	0.6163	1.7447	1.1026
WS 21-2	200	300	173.9	0.19758	10.31579	10.295	1	0.011601	31.499	31.5	199.6	0.6654	1.7421	1.2477
WS 21-4	200	300	150	0.225	13.596	13.26	0.98	0.006373	38.418	31.5	202.1	0.6735	2.3794	1.2507
Average	200	300	158	0.19086	12.17386	11.74833	0.97	0.008159	35.854	31.5	195.5	0.6518	1.9554	1.2033
					Fully graded large aggregate specimens									
WS 22-1	250	450	225	0.2048	15.3889	10.625	0.69	0.005084	37.085	29.1	303.4	0.6742	1.7642	0.6546
WS 22-2	250	450	259	0.316	16.23485	10.7843	0.66	0.00817	33.762	29.1	321.8	0.7152	2.2657	0.8417
WS 22-4	250	450	225	0.2227	15.8898	12.6977	0.8	0.005536	34.056	29.1	301.1	0.669	1.7808	0.7823
Average	250	450	236	0.247833	15.83785	11.369	0.72	0.006263	34.968		308.8	0.6861	1.9369	0.7595
WS 23-2	250	600	300	0.53516	23.4065	16.52465	0.71	0.007993	23.136	29.1	412.2	0.6871	2.4653	0.8817
WS 23-4	250	600	300	0.56419	24.3396	17.451	0.72	0.009761	18.945	29.1	395.3	0.6588	2.2596	0.9311
Average	250	600	300	0.549675	23.87305	16.98783	0.715	0.008877	21.041	29.1	403.8	0.6729	2.36245	0.9064
WS 24-1	250	800	400	0.41888	36.8333	29.42466	0.8	0.005019	36.31	29.1	520.9	0.6511	2.8669	1.3595
WS 24-3	250	800	400	0.34477	35.8587	31.575	0.88	0.005382	33.86	29.1	489.1	0.6113	2.3851	1.459
WS 24-4	250	800	400	0.43085	32.1778	19.594	0.61	0.005098	35.744	29.1	539.3	0.6741	2.7664	0.9054
Average	250	800	400	0.398167	34.9566	26.86455	0.76	0.005166	35.305	29.1	516.4	0.6455	2.6728	1.2413
WS 25-1	250	1000	500	0.5645	43.7109	25	0.57	0.006405	28.198	29.1	631.9	0.6319	2.8141	1.0332
WS 25-2	250	1000	500	0.55025	45.1633	34.093	0.75	0.005643	32.009	29.1	643.2	0.6432	3.0427	1.409
WS 25-4	250	1000	500	0.32687	39.194	20.5898	0.53	0.004739	38.116	29.1	608.6	0.6086	2.308	0.8509
Average	250	1000	500	0.48054	42.6894	26.56093	0.62	0.005596	32.77	29.1	627.9	0.6279	2.7216	1.0977
WS 26-1	250	1200	600	0.6944	54	45.168	0.84	0.007195	24.956	29.1	733.5	0.6112	2.9314	1.7041

13.5 Discussion on Wedge-splitting Test Results

continue

Specimen	B (mm)	D (mm)	a_0 (mm)	$CMOD_c$ (mm)	P_{max} (kN)	P_{ini} (kN)	P_{max}/P_{ini}	C_i (m/N)	E (GPa) (7)	E (GPa) Tested value	a_c (mm) (8)	γ_c	K_{IC}^{un} (MPa·m$^{1/2}$) (9)	K_{IC}^{ini} (MPa·m$^{1/2}$) (9)
WS 26-2	250	1200	600	0.65187	47.3404	37.663	0.8	0.005229	34.338	29.1	809.2	0.6743	3.3256	1.4209
Average	250	1200	600	0.6731	50.6704	41.4155	0.82	0.006212	29.647	29.1	771.35	0.6428	3.1285	1.5625
						Wet-screening specimens								
WS 32-1	200	300	150	0.149677	11.22078	8.23376	0.73	0.0075	32.643	29.1	183.4	0.6114	1.5241	0.7766
WS 32-2	200	300	150	0.12	9.4329	8.8478	0.94	0.007594	32.239	29.1	180.2	0.6008	1.232	0.8345
WS 32-3	200	300	150	0.1063	10.72727	8.431034	0.79	0.006118	40.014	29.1	178.4	0.5947	1.3705	0.7952
Average	200	300	150	0.125326	10.46032	8.504198	0.82	0.007071	34.96	29.1	180.7	0.6023	1.3755	0.8021
WS 33-2	200	600	300	0.22745	24.51087	18.973	0.77	0.006928	33.364	29.1	335.2	0.5586	1.9571	1.2654
WS 33-3	200	600	300	0.27451	21.9565	18.58836	0.85	0.006885	33.574	29.1	368.6	0.6144	2.1321	1.2397
WS 33-4	200	600	300	0.2389	21.2419	16.09071	0.76	0.006462	35.77	29.1	364.2	0.6069	2.0062	1.0731
Average	200	600	300	0.246953	22.56976	17.88402	0.79	0.006758	34.24	29.1	356	0.5933	2.0318	1.1927
WS 34-1	200	800	400	0.25049	27.3491	22.30603	0.82	0.006176	36.878	29.1估计	462.2	0.5777	2.0162	1.2884
WS 34-2	200	800	400	0.28242	27.0492	18.7662	0.69	0.006469	35.212	29.1	474.6	0.5933	2.1055	1.0839
WS 34-4	200	800	400	0.4157	32	26.7106	0.83	0.007059	32.269	29.1	493.2	0.6165	2.7126	1.5428
Average	200	800	400	0.316203	28.79943	22.59428	0.78	0.006568	34.79	29.1	476.7	0.5958	2.2781	1.305
WS 35-2	200	1000	500	0.27141	29.866	21.337	0.71	0.005664	39.858	29.1	592.1	0.5921	2.0705	1.1023
WS 35-3	200	1000	500	0.3281	25.634	19.2842	0.75	0.007534	29.965	29.1	602.3	0.6023	1.844	0.9962
WS 35-4	200	1000	500	0.3817	32.7	21.837	0.67	0.006945	32.51	29.1	600.5	0.6005	2.3362	1.1281
Average	200	1000	500	0.32707	29.4	20.8194	0.71	0.006714	34.11	29.1	598.3	0.5983	2.0836	1.0755

specimen should be set at the Quartern point of width of the specimen. The width and height of the specimen are $2H$ and D, respectively. G and P_{AL} are specimen dead-weight and the support reaction, respectively. $P_{AL}=G/2+P_V$. When the crack length is a, the force moment generated by external force to the crack tip can be expressed as follows:

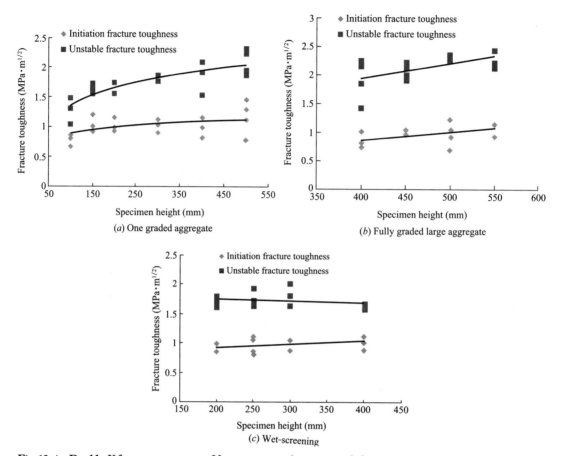

Fig. 13-4 Double-K fracture parameters of dam concrete with various graded aggregates tested from TPB specimens

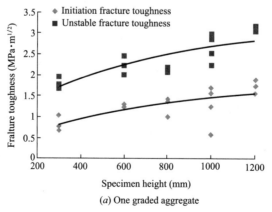

Fig. 13-5 Double-K fracture parameters of dam concrete with various graded aggregates tested from WS specimens (1)

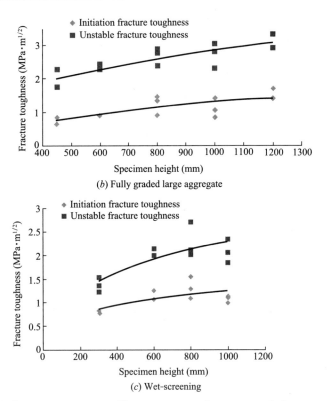

Fig. 13-5　Double-K fracture parameters of dam concrete with various graded aggregates tested from WS specimens (2)

$$M = P_H a + P_V(65 - H/2) \quad (13\text{-}10)$$

Let $M_H = P_H a$, $M_V = P_V(65 - H/2)$, thus equation (13-10) can be expressed as:

$$M = M_H + M_V \quad (13\text{-}11)$$

To the specimens with the same boundary condition and geometrical condition, when the specimen is compact tension (CT) loaded, the force moment at the crack tip is expressed by:

$$M = P_H a - M_H \quad (13\text{-}12)$$

When $2H = 270$mm, the $M_V = 0$. For the WS specimen and CT specimen with the same boundary condition and geometrical condition, it can be considered that the loading mode has a little effect on the crack tip stress field. In such a case, the fracture parameter formula of WS specimen can adopt that of CT specimen. If $2H > 270$mm, thus $M_V < 0$. It will lead to an effect of crack closure. And the wider is the specimen width, the more significant is the closure effect. If $2H < 270$mm, then $M_V > 0$. It will lead to an effect of crack opening. So, if the fracture parameter formulae of WS specimen still adopt that of CT specimen, it is not perfect in principle.

As illustrated in Fig. 13-6(b), it is the WS testing method recommend by the Norm. Take a WS specimen which $2H = 200$mm as an example, the additional bending moment $M_V = 15 P_H \tan 15° \approx 4.02 P_H$ and the effective bending moment $M_H = 80 P_H$, When the initial cracking occurs (a_0=80mm), $M_V \approx 5\% M_H$ can be obtained. In this condition, the WS load is less than the corresponding CT load which results from the additional bending moment M_V. It results in the corresponding smaller calculated fracture toughness.

In addition, if support is a single support, the dead-weight G and P_v all contributes to the yielding of positive moment at the crack tip. It will decrease the load and result in a lower calculated fracture toughness.

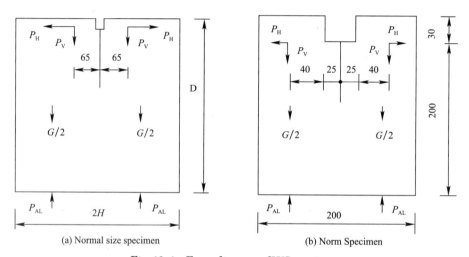

(a) Normal size specimen   (b) Norm Specimen

Fig. 13-6 Force diagram of WS specimen

13.6 Size Effect and Geometry Effect of Fracture Toughness

13.6.1 Size effect of fracture toughness

It can be seen from Fig. 13-4 that the double-K fracture parameters of TPB specimens tended to be size-independent. They can be regarded as the material constant.

Fig. 13-5 shows that the double-K fracture parameters of WS specimens existed certain size effect. The double-K fracture parameters became larger with increase of specimen width.

However, if $2H > 270$mm, then $M_V < 0$. It will result in the effect of crack closure. The wider is the specimen width, the more significant is the closure effect. It caused the fracture load and the calculated fracture toughness to increase. This analysis explains the reason why the double-K fracture parameters of WS specimens existed size effect in Fig. 13-6.

Literature[11] collected the compact tension test data conducted by various scholars all over the world. The results showed that the double-K fracture parameters obtained by these CT test data were size-independent. Therefore, the size-independent fracture parameters may be obtained if a countermeasure can be taken to eliminate the impact of M_V on the stress field at crack tip. This conclusion still needs further investigation.

13.6.2 Geometry effect of fracture toughness

To study the specimen geometry effect, TPB specimen and WS specimen which has the same specimen height D were selected. The TPB specimen and WS specimen with the same mix proportions and specimen height D for comparative study are SL1 and WS12、SL5 and WS13、SL8 and WS18、SL9 and WS19、SL10 and WS20、SL11 and WS21、SL44 and WS22, SL49 and WS32, respectively. The comparison of double-K fracture parameters of these two geometries specimens are listed in Table 13-6. For the transmitting force device employed in the Norm, when the specimen width was close 270mm, it resulted in the M_v being very small and the effect on fracture parameters being less.

Geometry effect of double-K fracture parameters Table 13-6

Specimen	Specimen height (mm)	Aggregate size d_{max}(mm)	Cubic compressive strength f_{cu} (MPa)	Unstable fracture toughness (MPa·m$^{1/2}$) (3)	Unstable fracture toughness (MPa·m$^{1/2}$) (5)	Initiation fracture toughness (MPa·m$^{1/2}$) (3)	Initiation fracture toughness (MPa·m$^{1/2}$) (5)	Ratio of unstable fracture toughness (3)	Ratio of unstable fracture toughness (5)	Ratio of initiation fracture toughness (3)	Ratio of initiation fracture toughness (5)
SL1	300	10	43.8	1.3626	1.4518	0.5889	0.6474	1.02	1.08	1.02	1.12
WS12	300	10	43.8	1.3381		0.5788					
SL5	300	20	43.3	1.7028	1.7869	0.96	1.0184	0.95	0.99	1.17	1.25
WS13	300	20	43.4	1.8023		0.8176					
SL8	300	20	53.7	1.8726	1.9592	1.0172	1.0756	1.04	1.09	0.85	0.90
WS18	300	20	48.1	1.7961		1.1941					
SL9	300	20	56.4	1.7331	1.8074	0.7672	0.8256	1.12	1.17	0.92	0.99
WS19	300	20	56.4	1.5427		0.8313					
SL10	300	20	50.2	1.9339	2.0194	0.9058	0.9641	1.03	1.08	0.79	0.84
WS20	300	20	50.2	1.8713		1.1465					
SL11	300	40	51.2	1.8258	1.9291	0.8529	0.92	0.93	0.99	0.71	0.77
WS21	300	40	50.4	1.9554		1.2003					
SL44	450	80	47.1	1.8831	2.0579	0.886	0.9935	0.97	1.06	1.17	1.31
WS22	450	80	/	1.9369		0.7595					
SL49	300	40	47.1	1.7384	1.8209	0.9343	0.9926	1.26	1.32	1.16	1.24
WS32	300	40	40.8	1.3755		0.8021					

The unstable fracture toughness ratio of the TPB specimen to the WS specimen by employing formula (13-3) fell within the range of 0.93-1.12, and the initiation fracture toughness ratio fell within the range of 0.71-1.17. If the formula (13-5) was adopted, the above-mentioned values fell within the range of 0.99-1.17 and 0.77-1.31, respectively. The various factors had great effect on the initial cracking, which resulted in the difference of initial toughness between these two kinds of specimens being significant. However, the unstable fracture toughness fell in a narrow dispersion band.

In general speaking, the different geometries of the specimens with different boundary conditions have influence on the stress field around the crack tip. The effects will bring difference of fracture parameters. However, viewed from test accuracy and practical engineering, as long as the mechanical conditions including small scale yielding and plane strain were met, the measured fracture toughness were almost equal by the different geometries specimens. Namely, fracture toughness is independent of specimen geometry and only material constant.

13.7 Conclusions

The size effect and geometry effect of double-K fracture toughness were studied by TPB method and WS method. The initial cracking load was measured by resistance strain gauges fixed along the path on both sides of the ligament. The following conclusions can be drawn:

(1) It is feasible to measure the initial cracking load by using resistance strain gauge. $P_{ini}/P_{max}=0.6\sim0.85$.

(2) The double-K fracture parameters of fully graded large aggregate concrete were greater than those of one graded aggregate concrete which has the same strength as the fully graded concrete. The double-K fracture parameters of wet-screening two graded aggregate concrete were almost the same as those of the one graded aggregate concrete.

(3) The double-K fracture parameters of concrete obtained by TPB tests are size-independent.

(4) For WS specimen, the effect of the additional moment M_V on the stress field of crack tip resulted in the size effect of concrete fracture toughness. When the effect was eliminated or reduced to some extent, the double-K fracture parameters obtained by WS tests also tended to be size-independent.

(5) The unstable fracture toughness determined by TPB and WS specimen which has the same mix proportions and specimen height lay in a narrow dispersion band and existed no obvious geometry effect. However, the effect of various factors on initial cracking of concrete was significant, thus the resulted difference of initial fracture toughness between tested by TPB and that by WS specimen couldn't be ignored.

(6) From a view of test accuracy and practical engineering, as long as the mechanical conditions including small scale yielding and plane strain were met, the double-K fracture parameters are only material parameters and independent of specimen geometry and size.

(7) For the WS testing method in Norm, the additional bending moment will exist around the crack tip. Even if the effect is smaller, the countermeasure which adjust transmitting force device to eliminate the additional moment is still recommended for the sake of improving the Norm.

References

[1] Mindess S.. Fracture Process Zone Detection, Fracture Mechanics Test Method for Concrete [R]. Report of Technical Committee 89-FMT, RILEM, Chapman and Hall, 1991.

[2] Xu, Shilang and Zhao, Guofan. The investigation on the propagation process of a crack in the concrete by means of photoelastic coatings [J]. Journal of Hydroelectric Engineering, 1991, 10 (3): 8-18.

[3] Xu, Shilang and Zhao, Guofan. The stable propagation of crack in concrete and the determination of critical crack tip opening displacement [J]. Journal of Hydraulic Engineering, 1989, 20 (4): 33-44.

[4] Hillerborg, A., Modeer, M. and Petersson, P.E.. Analysis of crack formation and crack growth in concrete by means of fracture mechanics and finite elements [J]. Cement and Concrete Research, 1976, 6: 773-782.

[5] Bazant Z P, Oh B H. Crack band theory for fracture of concrete [J]. Materials and Structures, 1983, 16 (93): 155-177.

[6] Jenq Y S, Shah S P. Two parameter fracture model for concrete [J]. Journal of Engineering Mechanics, 1985, 111 (10) : 1227-1241.

[7] Swartz S E, Refai T M E. Influence of size on opening mode fracture parameters for precracked concrete beams in bending[C]// Proceedings of SEM-RILEM International Conference on Fracture of Concrete and Rock. Houston, 1987: 242-254.

[8] Karihaloo B L, Nallathambi P. Effective Crack Model for the Determination of Fracture Toughness (K_{IC}s) of Concrete [J]. Engineering Fracture Mechanics, 1990, 35 (4/5) : 637-645.

[9] Bazant Z P, Kim J K, Pfeiffer P A. Determination of fracture properties from size effect tests [J]. Journal of Structural Engineering (ASCE) , 1986, 112 (2) : 289-307.

[10] Xu, Shilang , Zhao, Guofan. Double-K fracture criterion for crack propagation in concrete structures China Civil Engineering Journal, 1992, 25 (2) : 32-38.

[11] Xu Shilang, Reinhardt H W. Determination of double-K criterion for crack propagation in quasi- brittle fracture, PartI, Part II and Part III [J]. International Journal of Fracture, 1999, 98 (2) :111-193.

[12] Xu Shilang, Reinhardt H W. Crack extension resistance and fracture properties of quasi-brittle softening materials like concrete based on the complete process of fracture [J] .International Journal of Fracture, 1998, 92: 71-99.

[13] RILEM Technical Committee 50-FMC. Determination of the fracture energy of mortar and concrete by means of tree-point bend tests of notched beams [J] . Materials and Structures, 1985, 18 (106) :285-296.

[14] Nordtest Method. NT BUILD 491, Concrete and Mortar, Hardened: Fracture Energy (Model I)-Three-point bend tests on notched beams [S] . 1999.

[15] Japan Concrete Institute. JCI-TC 992, Test method for fracture energy of plain concrete (Draft) [S] . 2001J.

[16] CEB-Comite Euro-International du Beton. CEB-FIP Model Code [S]. Lausanne Bulletin D'Information No.213/214, 1990.

[17] Electric Power Industry Standard of the P. R. China (DL/T 5332-2005). Norm for fracture test of hydraulic concrete [S]. Beijing: China Electric Power Press, 2006.

[18] National Standard of the P. R. China (GB4161-84). Metallic Materials Plane Strain Fracture Toughness K_{IC} Test Methods [S]. Beijing: Standards Press of China, 1997.

14 Analysis and Research on Crack Prevention in Hole Roof for the Three Gorges Project

14.1 Presentation of the Problem

The Three Gorges Project is the largest multipurpose water control project under construction in the world nowadays. Its main works includes dam, powerhouse and navigation structure. The dam is of concrete solid gravity type, and the powerhouse, which is arranged in both banks, is one structure at the dam toe. Various holes and galleries, which are arranged dispersively and constitute important parts of the main works of TGP, include bottom outlet and mid-level outlet of overflow dam section in river channel, different galleries inside the dam body such as curtain, drainage, observation, cable and access galleries, intakes and draft-tubes of generating units in the powerhouse, filling and discharging galleries of the navigation structure, outlet gallery of the floor and galleries for various specialized-purpose.

These holes and galleries not only cause complication of engineering structure, inconvenient for construction and design, but also constitute weak parts of the engineering structure, resulting in frequent occurrence of cracking, erosion and damage. Practice of construction of the Gezhouba Project showed that, though placement with concrete of high strength class of C35~C45 around them, cracks appeared in most key holes and galleries such as discharge outlet of the water release structure, bottom desilting outlet, floating log outlet, intake of the generating unit, draft-tube and water conveyance gallery of the shiplock. Cracks occurred on the floor and two sides of the holes or galleries appear relatively homogeneous in density and character. However, cracks appeared in the hole roof location look like very serious. In addition to the Gezhouba Project, similar phenomenon was found also in many other hydroelectric projects such as the Dongjiang River Station and Fengtan Water Power Station in Hunan province and Fengshuba Water Power Station in Guangdong province. Hence, for the world-level large-sized project, the Three Gorges Project, it's very necessary to conduct crack-preventing analysis and study for hole roof locations. For that purpose, the above-mentioned problem is studied and discussed in this chapter from the aspect of combination of design with construction.

14.2 Causes and Analysis

Reasons why cracks or crazes appear in the concrete of the hole roof structure are very complicated. General investigations showed that its main reason resides in raw material (such as quality of cement, defects of sand-stone aggregates), design (such as error of designed calculating model, impropriety in choice of allowable index, error of computing standard), construction (low accuracy in weighing, poor technology, inadequate curing and protection, formwork deformation) and some of objective factors such as sudden drop in temperature, water conveyance during construction, etc. In addition to the above-mentioned reasons, there exist some other reasons in combination of design and construction, which are usually neglected by people.

14.2.1 Paying no more attention to the construction procedure in designing

Generally, in making structural calculation, designers create model on the supposition that the entire structure has been completed, paying no attention to the progressive rise of the structure during in-situ construction, in other words, omitting influence of accumulative self-weight load on construction. This method of treatment, especially for the hole roof structure, will bring about a fairly large error of modeling.

When model is constructed according to the entire structure, influence of holes or galleries on the structure itself is not so remarkable as a result of the fact that the boundary dimension of the entire structure is much larger than that of the hole structure. So for the hole structure, calculation of the concrete strength class and reinforcement arrangement may be performed following the equation (14-1) below:

$$\sigma = \sigma_w + \sigma_1 + \sigma_t + \cdots = \sum \sigma_i < [\sigma] \tag{14-1}$$

Where σ_w represents dead-weight stress; σ_1 represents external load stress; σ_t represents temperature stress; $[\sigma]$ represents permissible stress.

However, when accumulated dead-weight load of construction is taken into consideration, the first lift for the hole roof will form a upper plate, regarded as a beam with height (h) equal to the lift thickness (h_1) in case of unit width ($B=1m$). Upon completion of construction of the first lift, the second lift will be an external load to the first lift and the third lift will be an external load to both the first and second lifts, and so on and so forth.

Taking construction of the second lift (height h_2) for example, its model for computation may be simplified as a result of the fact that the limited part of the first lift extended into the two ends can't constitute rigid ends. Its applied load is as follows:

$$q = \sum_{i=1}^{n} q_i$$

$q_1 = \gamma \cdot h_1$ (dead-weight load)

$q_2 = \gamma \cdot h_2$ (second lift load)

$q_3 = q_c$ (construction load) $\tag{14-2}$

$q_4 = q_e$ (designed external load)

$q_5 = q_t$ (temperature load)

Upon having taken the accumulated self-weight load of construction into consideration, the calculated results show that the reinforcements on the hole roof would be more dense and the grade of concrete strength has to be raised. To our surprised, execution of construction according to the calculated results produces very little effect and cracks appear still. This inspires us that we should scratch where it itches and starts with the prerequisite of causing cracks.

In fact, cracks in the hole roof originate mostly from the comprehensive effect of the following conditions:

(1) Tensile (breaking) strength of reinforcement is much lower than that of concrete, namely,

$$[\sigma_c] \ll [\sigma_s] \tag{14-3}$$

It means that the concrete has cracked before the reinforcement works, it would be of no use to arrange more reinforcements, and once the concrete cracks, its internal reinforcements will be corroded by rust, leading to a worse condition.

(2) The strain of concrete, which result directly in cracks of concrete, is the sum of self-volumetric strain ε_v, temperature strain ε_t and strain ε_p arising from external loads including accumulated self-weight, i.e.

$$\varepsilon_c = \varepsilon_v + \varepsilon_t + \varepsilon_p \tag{14-4}$$

According to the above-mentioned equation (14-4), it's found that using concrete of higher strength class will increase its value of ε_v accordingly, though its resistance against the stress σ is improved. Meanwhile, to offset the increment $\Delta\varepsilon_v$ due to increase in grade of strength, some part of the strength of concrete has to be worn off accordingly. In other words, increasing grade of concrete strength further more will not actually give a satisfactory result in crack-prevention.

(3) When starting construction of the second lift, the first lift has only one stage of 7-10 days. Under the pressure of loads of q_2 and q_3, bottom of the first lift (hole roof surface), which is highly susceptible to cracking at this period, will largely produce cracks, which will spread and even form cracks. At this time loads applied on the first lift include its self-weight, accumulated dead weight and construction load, etc. The structure mainly bears bending and shearing forces.

Since a large quantity of reinforcements are arranged on the hole roof, issue on open type (type I) concrete breaking may be taken into consideration, without consideration of slide-off type (type II). It's assumed that one of the cracks across central part of the hole roof is shown as figure 14-2. Considering that bending moment is the major object to be studied in the aforementioned type I issue, load q applied on the beam may be transformed into a pair of force couple M_q. According to the calculating model, the value of moment M can be calculated by use of method of moments. Finally, on basis of the concrete breaking theory, the stress strength factor K_1 for the type I issue can be obtained with the aid of boundary distribution method. The calculating formula of K_1 is as follows:

$$K_I = \sigma_b \sqrt{\pi a} \frac{1}{\sqrt{\pi}} \left[1.99 - 2.47\left(\frac{a}{h_1}\right) + 12.97\left(\frac{a}{h_1}\right)^2 - 32.17\left(\frac{a}{h_1}\right)^3 + 24.80\left(\frac{a}{h_1}\right)^4 \right] \quad (14\text{-}5)$$

Where h_1 stands for thickness of the first lift; a stands for depth of the crack; σ_b stands for the maximum bending stress applied on the cross section of the lift where cracks probably appear afterwards, and

$$\sigma_b = \frac{M}{\frac{1}{6}h_1^2} = \frac{6M}{h_1^2} \quad (14\text{-}6)$$

In the equation (14-5), the n-th power of (a/h_1) may be omitted, for the thickness of the first lift (h_1) is much higher than the depth of crack (a). Namely, we have:

$$K_I = \sigma_b \sqrt{\pi a} \frac{1.99}{\sqrt{\pi}} = 1.12 \sigma_b \sqrt{\pi a} \quad (14\text{-}7)$$

The equation (14-7) is the expression formula of K_1 to the type I issue. The fracture toughness of concrete material (K_{IC}) may be described as:

$$K_{IC} = \sqrt{2E\gamma} \quad (14\text{-}8)$$

Where
$E' = E/(1-v^2)$, representing a parameter related to the Young's modulus (E) and the Poisson's ratio (v); γ represents unit surface energy or surface tension.

For any given material, E, v and γ are given, so concrete fracture criteria may be established as follows:

$$K_1 \begin{cases} < K_{IC} \text{ (no cracking or stable state)} \\ = K_{IC} \text{ (critical state)} \\ > K_{IC} \text{ (unstable state)} \end{cases} \quad (14\text{-}9)$$

Previous practical experiences of construction showed that time interval when cracks appear on the hole roof was generally between the construction periods of the second and the third lifts, and then the cracks gradually tend to stability. Hence, combined effect of the above-mentioned factors, especially the accumulated loads and concrete self-strain have a significant influence on the occurrence of cracks on the hole roof.

To sum up, when performing crack-preventing design for the hole roof, full attention should be paid to the combined effects of the above-mentioned factors and feasible measures should be adopted to prevent from cracking.

14.2.2 Ignorance of design intention in construction

Generally, While carrying out construction of the hole roof of structures, contractor(s), from the angle of construction, often place priority to method of concrete placement, topographic condition of blocs and convenience of placement, and do not strictly comply with requirements of design as they do in the construction of key item of works. This is also an important factor resulting in occurrence of cracks on the hole roof.

(1) Improper order of concrete placement blocs

During placement of concrete for the second lift and above on the hole roof, the following order of placement is generally adopted by contractors:

$$D_{cs} = \begin{cases} S \to C \to S & \text{(placement in orders)} \\ S_{2i-1} + S_{2j-1} \to S_{2i} + S_{2j} & \text{(skipping placement)} \\ C \to S \to S & \text{(central part first and then the two ends)} \end{cases} \quad (14\text{-}10)$$

Where C represents central blocs, located over the hole roof; S represents blocs on both ends above the hole roof; i represents any bloc on both ends ($i=1, 2, \cdots\cdots n$); j represents any bloc on central part ($j=1, 2, \cdots\cdots m$).

It is easy to find that the $S \to C \to S$ order will create end load, the $S_{2i-1} + C_{2j-1} \to S_{2i} + C_{2j}$ order will create dispersive load and the $C \to S \to S$ order will create the uniform distribution load on top beam first, all of which are beneficial to occurrence of cracks. Only the $S \to S \to C$ order will create end restraint on hole roof first and then horizontal compressive stress, which is applied on the lastly placed parts of the hole roof, having active effect of crack prevention. However, this order of placement is rarely adopted by contractors owing to its inconvenience in field construction.

(2) Lack of perfection in construction technology

In placement of concrete on the hole roof, dense reinforcements and accumulation of dumped materials increase difficulty of construction and bring about various problems on construction technology. Furthermore, inadequate construction quality control will cause appearance of weak surface of concrete on the hole roof, leading to cracking of concrete.

(3) Inadequate curing after concreting

It includes mainly formwork removal head of time, insufficient or non-uniform curing, improper measures of temperature and moisture preservation and protection. Formwork removal head of time leave the concrete of the hole roof to bear load excessively early, insufficient or non-uniform curing means that favorable conditions for normal growing of the concrete strength can't be satisfied, improper measures of temperature and moisture preservation and protection make the concrete lose ability to resist against cold wave hit, etc., leading finally to the cracking of concrete.

Hence, we're required, while performing construction of the opening roof of structures, to have full understanding of the design intention, grasp influencing factors and weak links, and take effective measures of crack-prevention from the angle of construction.

14.3 Crack Control Measures

To sum up, crack-preventing measures will be the most effective in case of close cooperation of design unit with construction unit. The following concrete measures may be taken:

(1) While making preparations for construction of the second lift, horizontal compressive stress should be applied to the concrete of the first lift on the hole roof in order to prevent from cracking of concrete. This horizontal compressive stress may be obtained by means of widening blocs on both ends of the hole roof and giving priority to placement of these blocs, or combining blocs through joint grouting to form rigid ends on both ends of the hole roof, or adding external horizontal compressive stress such as prestressed structure, anchored structure, etc. Value of the horizontal stress required by crack prevention may be calculated by use of hole span, lift thickness (h_1), concrete rupture toughness (K_{IC}) and other mechanical parameters, with a same direction as that of axial pressure of the first lift.

(2) Fiber concrete or other compound material concrete may be used, through testing, within a certain thickness (H_a) of the hole roof in order to form a structural surface layer with high crack resistant performance. Concrete of this layer is required to have lower self-volumetric strain ε_v and satisfy or exceed design requirements in strength grade, erosion-resisting and anti-friction properties.

(3) Construction procedure should be optimized, performing construction in layers and in blocs in compliance with design requirements. During placement of concrete, method of placement of whole bloc should be adopted as far as possible. In case of placement in blocs, placement order of $S \rightarrow S \rightarrow C$ shall be used, placing concrete in both ends first and then concrete over the opening roof, to prevent from cracking of concrete. For thickness of any lift, larger thickness of the first lift and smaller thickness of the successive lifts are adopted, i.e.

$$h_1 > h_{i+1} \quad (i=1, 2, \cdots\cdots) \tag{14-11}$$

To reduce the unfavorable effect of the accumulated load of construction. With the rise of lifts, the normal lift thickness may resume, when the accumulated self-weight load has no longer influence on h_1.

(4) Construction technology of concrete placement for the hole roof should be improved. The technologies include formwork setting, reinforcement erection, concrete placement, vibration and curing. In the course of designing, manufacturing, installation, removal of formworks, bearing capacity and rigidity of the bottom formwork may be strengthened to an appropriate degree, accuracy of manufacturing and installation should be improved, and time interval of formwork removal may be extended as far as possible, under permission of the construction cycle. Upon completion of concrete placement, curing and surface temperature preservation of concrete on the hole roof should be carried out carefully, to ensure normal growing of concrete strength and improve its ability to resist against cold wave hit.

14.4 Conclusions

Following conclusion can be drawn through the above-mentioned analysis:

(1) In view of the fact that the holes or galleries of the Three Gorges Project lay an important position in the whole project and their construction quality will directly influence that of the whole project, great importance should be attached to them by designing unit, construction units and other units related to the construction of the Project.

(2) Weak combination of design with construction leads to cracking of concrete and developing of cracks in the hole roof. Hence, designing unit and construction unit should cooperate closely in technology to perform effective crack control.

(3) Cracking problem of the hole roof of structures is one of problems which have not been resolved well in construction of hydraulic structure. In the course of construction of the Three Gorges Project, it's proposed, on the basis of full study of the crack prevention design and construction, to adopt new design method and construction measures to prevent from cracking of concrete, and spread and apply the achieved experiences.

References

[1] Zhou Hougui, A Summary of Quality Treatment Method for the Gezhouba Second Stage Project, No.9, 1989, Yangtze River.
[2] Liu Guangting and Zhang Fude, Study and Practice of Arch-abdomen Dam, Qinghua University Press, First Edition, July, 1996.
[3] Zheng Xiulin, Mechanical Properties of Material, Northwest Industry University press, First Edition, April, 1991.
[4] T. L. Anderson, Fracture Mechanics, CRC Press. Inc. Florida, US, 1991.
[5] G. C. Sih, Mechanics of Fracture Initiation and Propagation, Kluwer Academic publisher, Netherland, 1991.

15 Construction Technology and Practice of Dam without Cracks

15.1 Introduction

As everyone knows, the concrete dam is the main dam type and most commonly used in dam structures of the world. Cracking is a common phenomenon for concrete dam. The investigation report in 1988 of the International Committee on Large Dams (ICOLD) indicated that the majority of concrete dam built all over the world had cracks more or less. Up to now, a total of 243 dams have suffered catastrophic damage. In the 1980s, China Institute of Water Resources and Hydro-power made a survey of the cracks for 15 large scale concrete dam. The results show that the cracks are the most in Danjiangkou's 97m high slotted gravity dam, the total number is 3332. The cracks are the least in Zhexi 104m high head dam, the total number is 120. The Dam Monitoring Centre made an examination on 96 dams which belong to China Guodian Corporation in 1991. It revealed that nearly 70 dams had large-scale cracks. Cracks are common in concrete dam.

Many cases of which concrete dam failed to safely operate and yield economic returns, even wrecked can be found both in China and abroad. For example, for the Dworshka gravity dam in the United States, the depth and width of the abutment crack which occurred on the upstream face were 50m and 2.5mm, respectively, and the water amount of infiltration of the crack was 483L/s. The Chencun Dam in China appeared many cracks which kept it by low water-level operation during 1977-1979. The horizontal large cracks at 105m altitude of downstream face of the dam propagated significantly. The cracks with of arch crown position reached 1.39mm. The crack depth of 10 dam section on riverbed exceeded 5m. All these demonstrated that the dam was damaged seriously.

Therefore, construction of no-cracks concrete dam becomes a major problem for dam engineering. Based on the construction practice of concrete dam in phase one and phase two, the Three Gorges dam construction in Phase Three employed an innovative way which developed a new set of construction technology and the corresponding process to obtain no-cracks dam.

15.2 Dam Concrete Construction Technology

The Three Gorges dam (TGP) in phase three is a concrete gravity dam, including the desilting hole dam section, the 15th-20th dam section and the 21th-26th dam section at the right bank plant, and the 1th-7th non-overflow dam section at the right bank. The dam crest elevation was 185m, the minimum foundation surface elevation was 30m, and the maximum dam height was 155m. The width of upstream and downstream was 118m. The two longitudinal joints were set. The largest block was 52.5m×25m and the amount of cast concrete was more than 4000000m³.

15.2.1 Concrete design: optimize the mix proportions of concrete

Aimed at the high-grade concrete which was easy to crack, the countermeasures were adopted such as increasing the amount of fly ash, optimizing the admixture content and employing the low-heat cement and polycarboxylate superplasticizer in the mixture.

(1) Optimizing admixture content of the high-grade concrete

The high-grade concrete $R_{90}300F250W10$ which was used around the water intake of the dam in phase three appeared initial temperature rise going up too fast. Consequently, it was difficult for temperature control of the dam. To solve this problem, the content of retarding plasticizer was increased from 0.6% to 0.7%, the water consumption was reduced 2 kg, and the water-to-binder ratio and sand ratio was raised from 0.45 to 0.48 and from 28% to 30%, respectively based on the original dam concrete mix proportions. The adjusted $R_{90}300F250W10$ concrete can meet the designed requirements and 18kg cementitious material was reduced per cubic meter concrete. The $R_{90}300F250W10$ concrete mixture proportions before and after adjustment were listed in Table 15-1.

$R_{90}300F250W10$ concrete mix proportions before and after adjustment   Table 15-1

Concrete	Main parameters of mix proportions						Unit weight (kg/m³)					
	Unit water amount	Water-to-binder ratio	Sand ratio (%)	Fly ash content (%)	W.R. (%)	A.E. (0.1‰)	C	F.A.	S	Small gravel	Medium gravel	Large gravel
Before adjustment	98	0.45	28	20	0.6	3	174	44	588	378	378	756
After adjustment	96	0.48	30	20	0.7	2.5	160	40	637	371	371	743

(2) Employing the low-heat cement and polycarboxylate superplasticizer in mass concrete

With regard to pouring the pump concrete $R_{28}250$ in high-temperature season, the low-heat cement and updated polycarboxylate superplasticizer were used to reduce the cement amount and hydration temperature rise. Table 15-2 shows the test data of adiabatic temperature rise on site of moderate-heat cement and low-heat cement.

Test data of adiabatic temperature rise of concrete $R_{90}200$ employing   Table 15-2
moderate-heat cement and low-heat cement 42.5 grade   Unit: ℃

Cement type	Time (d)	0	1	2	3	4	5	6	7	10	14	19
Moderate-heat	Temperature	13.6	21.1	24.4	27.2	28.5	29.3	30	30.6	31.3	31.79	32.0
	Temperature rise		7.5	11.3	13.6	14.9	15.7	16.4	17	17.7	18.19	18.4
Low-heat	Temperature	13.8	19.7	22.6	24.2	25.1	26	26.7	27.5	29.3	31.07	32.2
	Temperature rise		5.9	8.8	9.4	11.3	12.2	12.9	13.7	15.5	17.27	18.4

The results show that the highest temperature of moderate-heat cement concrete and low-heat cement concrete reached 32.03℃ and 32.2℃ after the thermometer was fixed for 19 days, respectively. The highest temperature and corresponding time of these two kinds of cement concrete were basically the same. However, the temperature of low-heat cement concrete in first 10 days was 2-3℃ lower than that of moderate-heat cement concrete, and 14 days later they kept basically the same temperature. The early age low temperature of the low-heat cement concrete was beneficial to controlling the highest temperature and preventing thermal cracking.

The polycarboxylate superplasticizer was employed in the pump concrete $R_{28}250F250W10$ of the dam in Phase Three, which resulted in the cement amount reducing from 273 kilogram to 241kilogram. It was important for temperature control and crack prevention of concrete.

15.2.2 Concrete mixing: strict control of exit temperature

The exit temperature of concrete at the base restraint area of main structure of TGP was controlled below 7℃ and that of concrete at outside of the base restraint area was controlled within 7-9℃ in other seasons except the winter, respectively.

To strictly control the exit temperature, some advanced technologies such as secondary air cooling of aggregates, adding of slate ice and chilled water to mixtures were employed. By the secondary air cooling of aggregates, the internal temperature of super-large gravel, large gravel reached 6-8℃. Meanwhile, the temperature of cement was controlled at 60℃ when it is sent into the tank. The amount of sliced ice and cold water added into concrete mixture were determined according to the temperature of aggregate and the sand moisture after the secondary air cooling of aggregates. The amount of sliced ice was usually 30-60kg/m^3. Thus, the exit temperature of the concrete produced at altitude 150m mixing system was tested for 5618 times and the qualified rate was 99.6%. For this concrete mixing system up to Jan. 20, 2006, the raw material quality inspection of concrete can meet the standard of Three Gorges Project (TGPS) requirements, and the various indexes have reached the excellent level of production of concrete. The total qualified rate of the exit temperature of concrete was 98.0%.

15.2.3 Transportation of concrete: strengthening the insulation and cooling

The delivery system for the TGP dam concrete employed a layout called "one-machine-one-belt". During the transportation of concrete, some cooling measures such as the sunshading along the way, covering the insulation quilt on warehouse and spraying were used to prevent the pouring temperature to exceed standard. All these make sure that the pouring temperature can meet the design requirements.

(1) Insulation and cooling of the concrete delivery system

When the tower belt crane was used for pouring, the concrete was directly delivered from mixing plant into placement section through the feeding line. As the longest feeding line of concrete of TGP in phase three reached up to 1100m, a countermeasure which the upper belt was closed and the back of the lower belt was water-flushed to decrease the concrete temperature was adopted to prevent calories absorb of the pre-cooling concrete in transportation. Consequently, the countermeasure helped to reduce 2℃ or more temperature recovery of concrete on feeding line.

(2) Cooling of working unit of dam

The working unit of dam construction was cooled by sprayer installed on both sides of the working unit, which can form the fog layer in the upper area of poured working unit so as to reduce ambient temperature of the working unit. The spray can help to reduce temperature around the working unit surface 5-6℃ compared to the air temperature.

(3) Insulation of working unit

Before pouring, heat preservation quilt with over 2/3-3/4 surface area of working unit should be provided. After vibration of casting layer, heat-preserved quilt will be immediately covered for concrete insulation. Practice shows that, for the larger unreinforced or less reinforced dam block, implementation of large area or whole working unit heat insulation can help to stop the pouring temperature from exceeding standard.

15.2.4 Concrete pouring: continuous improvement of operation technology

The concrete pouring of third-phase TGP dam employed a construction scheme which mainly took the tower-belt machine and took the gantry crane as supplementary equipment. The maximum control height of the belt machine and the tower crane was 165 meters and 185 meters, respectively.

The concrete pouring involved preparation of forms, steel bars and embedded parts, the working unit processing, concrete feeding and leveling, vibrating, etc. According to the importance of cracking resistance of concrete for every process, the improvement and innovation of the construction technology were carried out for the whole construction process of the third-phase TGP dam.

(1) Concrete form

The forms are mainly leveling forms, supplemented by partial composite steel forms, wooden forms, precast concrete forms, to satisfy the construction needs and the requirements of concrete appearance.

The large-scale integral lifting forms and steel forms were used on a large scale in construction such as the corbel integral lifting forms, the shaft (cable well and ventilation hole) integral lifting forms, the sealing integral lifting forms and the set-shaped steel forms of gallery and drainage and so on.

The corbel integral lifting form was mainly composed of panel, operation platform and concrete balance counterweight block. The main technical points were as follows: (a) The concrete balance counterweight block was set to keep the operation platform level when the form was installed and to stop form from shaking after demoulding. (b) The operation platform was ingeniously designed to be L-type structure so that the insulation was conveniently conducted in time after the corbel concrete was demoulded.

(2) Treatment of joint surface

The construction joint surface was treated by a high-pressure water jet with pressure of 30-50MPa so that large size sands and small gravel can be visible on surface. This operation was conducted after 24-36 hours later since the finishing of working unit. Those partial joint surfaces that fail to meet the requirements should be artificially chiseled to make the surface rough.

For the working unit of concrete dam blocks of long intermission casting, the last casting layer on the top was poured with fiber reinforced concrete, followed by strict vibration and finishing technology, in order to avoid cracking of the dam placement section caused by long intermission.

(3) Concrete feeding, leveling, vibration

In the third-phase TGP dam concrete construction, the prepared concrete was fed to the working unit by feeding line and tower belt machine. The flat casting method was employed. The distribution mode of feeding concrete adopted the overlapping the distributed concrete resembled as fish scales.

After feeding concrete, leveling was adopted firstly and then vibration. The two operations were separated to avoid lack of vibration, missing vibration and excessive vibration (Three Vibrations for short).

The concrete vibration followed the leveling. The duration of vibration was determined the situation appeared which the aggregates no longer obviously subsided. The water and bubbles no longer escaped and started bleeding (about 20-30s).

When the surrounding area of working unit of dam construction was poured, second vibration should be used. When the last layer of concrete of each working unit was poured, accumulated aggregates shall be dispersed in time, the vibration should follow excluding the bleeding of working unit.

In order to prevent the "Three Vibrations", a vibration machine timing alarm was independently developed. It helped the vibration to automatically quantify vibration time, monitor vibration depth and count the operation data of working unit of dam construction, and effectively improved the quality of concrete placement and the fine level of construction.

15.2.5 Concrete care: fine implement of concrete curing

When the pouring and vibration of concrete is completed, it is just like a newborn baby and begins his life cycle. It should be given careful nursing or curing otherwise it will result in cracking even severe failures.

(1) Curing

During the third-phase of TGP dam construction, concrete curing was taken as an independent project, for which, long-term constant curing measures have been taken. Regarding the concrete pouring of each working unit (storehouse), watering was carried out in time after closing the working unit. The working unit surface curing should be conducted till pouring the upper layer, and the curing period of the permanent exposed surface was much longer than that required by a standard. As for the curing in summer, the automatic watering curing with rotating nozzles was adopted. The water flow of each water sprinkler is 12-15L/min. For places out of reach by the sprinklers, manual watering should be conducted. ϕ25mm plastic pipes were used for curing the upstream and downstream surface as well as the vertical and horizontal gap. After ϕ1mm hole was drilled every 20-30 centimeters on the plastic pipe, it was hung on the form or exposed bar for watering. In terms of curing in winter, watering was adopted for all after the working unit was covered with heat preservation quilts so as to make the concrete surface always keep moist. For partial long-interval woring unit, specially assigned persons should be available for watering or sprinkling curing, thus keeping the surface moist.

(2) Water cooling

Personalized Water Cooling was implemented in the third-phase TGP dam construction. Namely, the layout of cooling water pipes was determined based on temperature variation of the different grade concrete. The water flow was controlled based on the water temperature dynamic control during water inflowing and outflowing. It improved the quality and efficiency of water cooling.

Cooling water pipes were generally placed by 1.5m×2.0m or 2.0m×1.5m in the working unit. The arrangement of water pipes for concrete $R_{28}250$ or higher grade concrete should be denser by 1.5m×1.0m or 1.0m×1.5m. ϕ25 diameter black iron pipe was used for cooling. If 3m is upgraded with the layout of two-layer water pipes, plastic pipes can be used for the second layer.

The water cooling of the third-phase dam is divided to the initial, middle and later stages. The initial-stage water cooling is mainly to cut down the temperature peak of concrete at early age, decreases the highest temperature inside the mass concrete so as to control the temperature within the allowed design scope. By around 10-day initial water cooling, the concrete temperature can basically be reduced to 24-28℃. By the middle-stage water cooling, the interior temperature of concrete is mainly reduced to 20-22℃. It will be beneficial to reduce the temperature difference between interior and exterior of concrete dam and make it go through the winter safely. The purpose of the later-stage water cooling is to cool the parts which need to joint or joint grout so that the grouting temperature can reach to 14-17℃.

① Initial-stage water cooling

Initial-stage water cooling of concrete was carried out within 12hrs after the working unit closing, while the water cooling for high-grade concrete should start earlier when the working unit concrete begins to be poured. From April to November, due to the high temperature of river water (>15℃), 8-10℃ cooled water was generally used for the initial-stage water cooling. From December to March of next year, the water temperature was 11-15℃, the river water can directly be used for initial cooling. For the high-grade concrete, strengthening measures should be taken, for example, the flow may be increased to 40L/min, or the cooled water at 8-10℃ may be adopted.

The control standards of cooling water flow at the initial stage are as follows: In seasons with high temperature from May to September, the water flow for $R_{90}150$ concrete is 15-20L/min; The water flow in the first 4 days after working unit closing for the $R_{28}200$-$R_{90}200$ concrete is 30-40L/min, and that in the last 6 days is 18-20L/min; The water flow in the first 4 days after the working unit closing for high-grade concrete is 40-50L/min, and that in the last 6 days is 20-25L/min. In other seasons, the water flow for $R_{90}150$ concrete is 15-18L/min; The water flow in the first 4 days after the working unit closing for $R_{28}200$-$R_{90}200$ concrete is 25-35L/min, and that in the last 6 days is 15-20L/min; The water flow in the first 4 days after the working unit closing for high-grade concrete is 30-40L/min, and that in the last 6 days is 18-20L/min.

During the water cooling process, the water inlet and outlet direction should be changed once every other day. Through dynamic adjustment of the water flow, the water temperature difference between inlet and outlet should be controlled above 5℃. If it is smaller than 5℃, then the water flow should be reduced to its lower limit of the control standard.

② Middle-stage water cooling

At the end of August, the dam temperature of the typical dam sections for each elevation scope was checked. Combined with the temperature observed by the instrument inside the dam, the water cooling order of pipes was determined. First of all, water cooling was carried our for the concrete poured from May to August, followed by the concrete poured in April and September. In late November, the concrete temperature of all middle-stage water cooling parts should be reduced to 20-22℃, so as to make sure the dam can go through the winter smoothly.

During the middle-stage water cooling with river water, the outlet water temperature of the cooling water pipe should be checked first, and the water cooling can be carried out only when the outlet temperature is over 2℃ higher than the inlet temperature. In late October, the first wind cooling of the mixing system was stopped, only the second wind cooling run for concrete production. The 8-10℃ cold water produced from the first wind cooling workshop was used for the middle-stage water cooling. During this period, the water flow was controlled according to the temperature observed by the instrument in the dam. It was ensured that the cooling rate of concrete is no greater than 1℃/d.

The water flow was usually controlled at 15-20L/min, and the water direction was changed every 3 days. The water cooling termination standard for the middle-stage water cooling with 10℃ cold water was that the water temperature outlet was lower than 18℃. When the dam temperature was decreased to 20-22℃, thermal insulation should be conducted for 5 days.

③ Later-stage water cooling

Thermal insulation of concrete was carried out before the later-stage water cooling. According to the thermal insulation temperature, whether employing the river water or cooled water was determined. When the concrete block temperature exceeded the cooled water which was 15℃, the river water was used first for lowering the temperature. When the temperature difference between inlet and outlet reached within 3℃, the refrigeration water was employed to cooling it down to the grouting temperature.

(3) Heat preservation

The temperature difference of concrete between interior and exterior of dam should be controlled and reduced, thus a steady and uniform temperature field could be formed. It is the key to prevent thermal cracking of mass concrete. The TGP dam area has characteristics such as obvious seasonal variation, significant temperature difference between day and night, and frequent and sudden temperature drop. The lowest temperature of Three Gorges dam area in winter is −1.9℃, and its highest temperature in summer is 39.8℃. There are 8-10 times of temperature drops every year. Under such climatic conditions, heat preservation of concrete has to be enhanced in the winter, the spring and the autumn.

The polyethylene plastic insulation quilt was adopted by the temporary construction. The impermeable layer and long-interval construction surface employed 3cm-thick heat preservation quilt. The 2cm-thick heat preservation quilt was used for temporary heat preservation for other parts. For the permanent exposed surfaces of upper reaches and lower reaches, polystyrene board and external coating of waterproof paint were pasted on the exposed surface for insulation. For the area around water inlet hole, 2cm-thick polyurethane rigid foam was coated for insulation. For holes such as water inlet holes, the canvas was applied for sealing and prevention of wind from passing through it.

The polystyrene board insulation material employed on the upstream and downstream permanent exploded surface includes the binder, polystyrene board, and waterproof coating. The main physical and chemical properties of the polystyrene board are shown in Table 15-3. The tested temperature of 5cm-thick polystyrene board used on the upstream surface of 18-2 deck at the right bank workshop dam section is shown in Fig. 15-1.

Physical and chemical properties of the polystyrene board for heat preservation Table 15-3

Item	Apparent density (kg/m³)	Dimensional change rate (%)	Water absorption (%)	Compressive strength (MPa)	Thermal conductivity [W(m·K)⁻¹]	Water vapor penetration coefficient (ng/Pa·m·s)
Performance	>20	<5	≤4	≥0.10	0.034	≤4.5

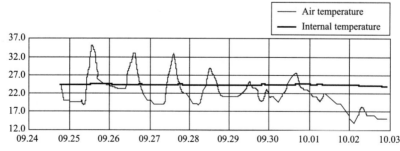

Fig. 15-1 Comparison between the tested temperature of the 5-cm-thick polystyrene board and the air temperature

The test results show that the air temperature ranged for 14.0℃ to 35.0℃, the change of temperature was 21.0℃, and the maximum change of temperature in 2hrs was 6.8℃. The internal temperature of the polystyrene board ranged from 23.8℃ to 24.4℃, the change of temperature was 0.6℃. Basically, the 5cm-thick polystyrene board was maintained a state of constant temperature and humidity.

The polyurethane rigid foam insulation material around the water inlet hole is composed of the main material and auxiliary material, including the foaming agent, catalyst, stabilizer and fire retardant. Before adding the foam agent, the polyurethane rigid foam is a strong adhesive. After adding the foam, its adhesion stress is still very strong to connect with concrete. In addition, this kind of material had a advantage of non-easy falling off after it is soaked in water. Its main physical and chemical properties are shown in Table 15-4.

Physical and chemical properties of the polyurethane rigid foam insulation material   Table 15-4

Item	Density (kg/m^3)		Dimensional change rate (%)	Water absorption (g/m^2)	Compressive strength (MPa)	Thermal conductivity [W(m·K)$^{-1}$]	Fire resistance (s)
	Internal density	Surface density					
Performance	29~60	35~50	<2.0	<150	>0.17	0.019±0.003	<3

The heat preservation of dam was carried ou at the end of September for each year. It was pasted 5cm-thick polystyrene board in the area of upstream surface below elevation 98m, 3cm-thick polystyrene board in the area of upstream surface over elevation 98m, and 3cm-thick polystyrene board in the area of downstream surface of dam.

15.3 Conclusions

Since July 2003, the third-phase TGP dam began its first concrete pouring, the dam concrete pouring work was completed successfully on May 20, 2006. A set of construction technology and process for construction of the third-phase TGP dam was developed by means of fine studies, improvement, testing and innovation from all aspects of construction.

The successful construction practice of the third-phase TGP dam demonstrated that the fine control of whole construction process is very important as the scientific design and advanced equipment to build a no cracks dam successfully.

References

[1] Three Gorges Project Construction Headquarter of the China Gezhouba Group Company. TGP/CI-3-1B construction report [R]. Yichang, Hubei Province, China, 2006.04.

[2] Zhou Hougui. Study on bringing forward the construction of dam and power house of the third-stage project in Three Gorges Project [J]. Red River, 2006, 25(1): 1-3.

[3] Zhou Jianhua, Teng DongHai, Zhang Hongbing. Prevention and control of cracks in the third-phase TGP dam concrete [J]. Gezhouba Group Science & Technology, 2006, (4):60-62.

[4] Zhou Hougui. New technology of concrete construction in Three Gorges Project [J]. Advances in Science and Technology of Water Resources, 2008, 28(2): 42-46.